Modern Bioenergy for Sustainable Development

Modern Bioenergy for Sustainable Development

Editors

Dilip Khatiwada
Pallav Purohit

MDPI • Basel • Beijing • Wuhan • Barcelona • Belgrade • Manchester • Tokyo • Cluj • Tianjin

Editors

Dilip Khatiwada
KTH Royal Institute of Technology
Sweden

Pallav Purohit
International Institute for Applied Systems Analysis (IIASA)
Austria

Editorial Office
MDPI
St. Alban-Anlage 66
4052 Basel, Switzerland

This is a reprint of articles from the Special Issue published online in the open access journal *Sustainability* (ISSN 2071-1050) (available at: https://www.mdpi.com/journal/sustainability/special_issues/ModernBioen_Poten_Strat).

For citation purposes, cite each article independently as indicated on the article page online and as indicated below:

LastName, A.A.; LastName, B.B.; LastName, C.C. Article Title. *Journal Name* **Year**, *Volume Number*, Page Range.

ISBN 978-3-0365-0468-1 (Hbk)
ISBN 978-3-0365-0469-8 (PDF)

Contents

About the Editors vii

Preface to "Modern Bioenergy for Sustainable Development" ix

Dilip Khatiwada and Pallav Purohit
Special Issue on Assessing the Modern Bioenergy Potential and Strategies for Sustainable Development: Transformations through Nexus, Policy, and Innovations
Reprinted from: *Sustainability* **2021**, *13*, 374, doi:10.3390/su13010374 1

Dilip Khatiwada, Pallav Purohit and Emmanuel Kofi Ackom
Mapping Bioenergy Supply and Demand in Selected Least Developed Countries (LDCs): Exploratory Assessment of Modern Bioenergy's Contribution to SDG7
Reprinted from: *Sustainability* **2019**, *11*, 7091, doi:10.3390/su11247091 7

Mngereza Miraji, Xi Li, Jie Liu and Chunmiao Zheng
Evaluation of Water and Energy Nexus in Wami Ruvu River Basin, Tanzania
Reprinted from: *Sustainability* **2019**, *11*, 3109, doi:10.3390/su11113109 37

Richmond Antwi-Bediako, Kei Otsuki, Annelies Zoomers and Aklilu Amsalu
Global Investment Failures and Transformations: A Review of Hyped *Jatropha* Spaces
Reprinted from: *Sustainability* **2019**, *11*, 3371, doi:10.3390/su11123371 49

Spyridon Achinas, Johan Horjus, Vasileios Achinas and Gerrit Jan Willem Euverink
A PESTLE Analysis of Biofuels Energy Industry in Europe
Reprinted from: *Sustainability* **2019**, *11*, 5981, doi:10.3390/su11215981 73

Andreia Marques Postal, Gabriela Benatti, Mar Palmeros Parada, Lotte Asveld, Patrícia Osseweijer and José Maria F. J. Da Silveira
The Role of Participation in the Responsible Innovation Framework for Biofuels Projects: Can It Be Assessed?
Reprinted from: *Sustainability* **2020**, *12*, 10581, doi:10.3390/su122410581 97

About the Editors

Dilip Khatiwada is an assistant professor at the Division of Energy Systems, Department of Energy Technology, KTH Royal Institute of Technology, Sweden. His research provides insights on the sustainability of renewable energy and its potential role in the mitigation of climate change, improved energy security, and sustainable development in different development contexts. He has also made methodological contributions for assessing sustainability in connection with energy and climate policies. Dr. Khatiwada's research interests include lifecycle assessment (LCA), energy systems optimization, nexus assessment, and bio-based economy. Dr. Khatiwada has around 20 years of work experience in several private, public, and international development agencies. He has worked as a visiting researcher at the Brazilian Bioethanol Science and Technology Laboratory (CTBE) in Brazil and spent research periods at the International Institute for Applied Systems Analysis (IIASA) in Austria and the Institute for Sustainability, Energy, and Environment, University of Illinois at Urbana-Champaign (UIUC), in the USA. He has also spent a research and teaching period at the Integrated Research System for Sustainability Science (IR3S), University of Tokyo, Japan. Dr. Khatiwada has worked as an under-secretary in the Government of Nepal and as a consultant at the Asian Development Bank. He is a mechanical engineer, and he also holds a master's degree in Sustainable Energy Engineering from KTH Royal Institute of Technology.

Pallav Purohit is a researcher in the Energy, Climate, and Environment (ECE) Program of the International Institute for Applied Systems Analysis (IIASA). He has developed and implemented the global fluorinated greenhouse gases (F-gas) module in IIASA's Greenhouse Gas - Air Pollution Interactions and Synergies (GAINS) model and coordinated various policy applications involving the GAINS model on the mitigation of air pollutants and greenhouse gases in industrialized and developing countries. Before joining IIASA, Dr. Purohit worked as an e8 Postdoctoral Research Fellow at the Hamburg Institute of International Economics (HWWI), Germany, where his focus was on a detailed technical evaluation of renewable energy options towards a more policy-oriented analysis of the chances and risks of such technologies under the Clean Development Mechanism of the Kyoto Protocol. He was also a visiting faculty member at the Institute of Political Science at the University of Zurich, Switzerland, and at the Department of Built Environment and Energy Technology at the Linnaeus University, Sweden, and a visiting fellow at the School of International Development, University of East Anglia, UK, and Council on Energy, Environment and Water (CEEW), New Delhi. He is an Associate Editor for the *International Journal of Global Energy Issues* and an Editorial Board member of *Sustainability, Innovative Energy and Research, and Current Environmental Management*, among other journals. Dr. Purohit received his MSc in Physics from the H.N.B. Garhwal University, India, in 1998 and his PhD in Energy Policy and Planning from the Indian Institute of Technology Delhi in 2005. He has contributed to several UNEP, IEA, World Bank, and IPCC reports.

Preface to "Modern Bioenergy for Sustainable Development"

Access to affordable, reliable, and clean energy is a key Sustainable Development Goal (SDG 7), which also underpins other SDGs since energy access facilitates economic development, food security, health and well-being, education, and other related objectives. A large share of global renewable energy is supplied by biomass; however, this is, in many cases, traditional biomass and waste use, particularly in low- and middle-income countries (LMIC), where biomass provides the main energy source for domestic and productive uses at different scales, mainly for generating heat and cooking purposes. However, in LMICs, biomass use is often unsustainable and inefficient, resulting in deforestation, soil degradation, and health risks from household air pollution that disproportionately affect women and children. Introducing modern bioenergy as a substitute for fossil fuel is one of the options for reducing carbon dioxide emissions while also satisfying its rapidly increasing energy demand, provided that bioenergy is produced sustainably. Bioenergy can be converted into a variety of energy carriers/services, such as heat, electricity, liquid biofuels (i.e., biodiesel, bioethanol, methanol, dimethyl ether (DME), and Fischer–Tropsch (FT) synfuels), or gaseous biofuels (i.e., biogas and hydrogen). This means that modern bioenergy has the possibility of replacing each type of fossil fuel. Bioelectricity can also provide a base-load and plays a key role in balancing the grid and providing storage options.

The papers presented in this book highlight the role of modern bioenergy in the global transition to sustainable energy and outline its key elements and their significance with respect to the environment and development. The reader will find papers that explore a key question: What are the synergies and trade-offs between bioenergy use and the SDGs? The nexus approach can be useful in understanding the synergies and trade-offs, capturing the interdependence between scarce resources (e.g., water, energy, land). It is also critically important to explore cross-cutting issues such as socio-political, techno-economic, legal, and environmental ones when it comes to the deployment of biofuels for sustainable development. Investment plans also need to be addressed with consideration for the transformations and complexities associated with biofuels. Innovation in the transition towards biofuel will only be possible if we consider a participatory approach that considers people's behavior and societal needs.

The objective of the four research papers and one review article selected for this volume is to capture a comprehensive understanding of major issues associated with the large-scale development and dissemination of modern bioenergy technologies/systems. Despite diversity and complexity, we aimed to find systematic commonalities and differences, identify best practices, and fill the most critical knowledge gaps. As a result, the volume will be of great interest to policy, legal, and impact assessment experts, as well as private companies, community organizations, and all those interested in devising sustainable development options for a resource-scarce and carbon-constrained world through economic growth and diversification. This Special Issue consists of four research papers and one review, contributed by experts in the field, on topics including (i) mapping bioenergy supply and demand in selected least developed countries; (ii) evaluation of the water and energy nexus in Wami Ruvu River Basin, Tanzania; (iii) global investment failures and transformations in modern biofuels; (iv) a socio-political, techno-economic, legal, and environmental analysis of the biofuel energy industry in Europe; and (v) the role of participation in the responsible innovation framework for biofuels projects. We hope that this work will be inspirational for further research on modern bioenergy and sustainable development.

Dilip Khatiwada, Pallav Purohit
Editors

sustainability MDPI

Editorial

Special Issue on Assessing the Modern Bioenergy Potential and Strategies for Sustainable Development: Transformations through Nexus, Policy, and Innovations

Dilip Khatiwada [1] and Pallav Purohit [2,*]

[1] Division of Energy Systems, Department of Energy Technology, KTH Royal Institute of Technology, Brinellvägen 68, SE-100 44 Stockholm, Sweden; dilip.khatiwada@energy.kth.se
[2] Air Quality and Greenhouse Gases (AIR) Program, International Institute for Applied Systems Analysis (IIASA), Schlossplatz 1, Laxenburg A-2361, Austria
* Correspondence: purohit@iiasa.ac.at

Received: 30 December 2020; Accepted: 31 December 2020; Published: 3 January 2021

This editorial aims to summarize the five scientific contributions that contributed to the Special Issue assessing the modern bioenergy potential and strategies for sustainable development, considering the several aspects, viz. biofuels–water nexus, policy analysis, complexities of biofuel investment plans/options, and participatory innovation framework.

1. Overview

Modern bioenergy is widely acknowledged as a potential substitute for fossil fuels to offset human dependence on fossil fuels for energy. It has the potential to make an important contribution towards sustainable development, from rural development and poverty alleviation to energy security and climate change mitigation. Modern bioenergy would play a key role in the sustainable transformation of energy systems, along with its decisive contribution in developing a low-carbon growth path while contributing to energy diversification and security [1,2]. Nevertheless, there are concerns and scientific debate on sustainable bioenergy, especially on land-use change (LUC), food security, and biodiversity loss [3]. In order to minimize the potential risks and challenges, we need to understand interdisciplinary dimensions such as government policies, financing schemes, stakeholders' engagement, and intersectoral linkages while devising sustainable bioenergy strategies. Bioenergy accounts for the largest share of the global energy supply compared to all renewable energy resources [4]. In spite of the largest share among other renewables with 10% of the global primary energy supply [5], the current rate of bioenergy deployment is well below the levels required for low-carbon scenarios [6]. Accelerated deployment is urgently needed in order to ramp up the contribution of sustainable bioenergy across all sectors, notably in the transport sector, where consumption is estimated to triple by 2030. There is an increasing understanding that only bioenergy that is supplied and used sustainably has a place in a low-carbon energy future. Modern bioenergy has several positive benefits, and synergies need to be established between energy and food security through an integrated assessment of resources.

There is a plethora of bioenergy research either in the supply chains, conversion technologies, or techno-economic analysis of different biofuel production pathways [7]. However, little attention has been shown to how bioenergy production and consumption would have a role in sustainable development. What are the synergies and trade-offs between bioenergy use and the SDGs? The nexus approach can be useful in understanding the synergies and trade-offs, capturing the interdependence between scarce resources (e.g., water, energy, land). It is also critically important to explore cross-cutting issues such as socio-political, techno-economic, legal, and environmental ones when it comes to the deployment of biofuels for sustainable development. Investment plans also need to be addressed considering transformations and complexities associated with biofuels. Innovation in the transition towards biofuel would only be possible if we consider a participatory approach considering people's behavior and societal needs.

2. A Short Review of the Contributions to This Special Issue

This Special Issue consists of four research papers and one review, contributed by experts in the field, on topics including (i) mapping bioenergy supply and demand in selected least developed countries; (ii) evaluation of water and energy nexus in Wami Ruvu River Basin, Tanzania; (iii) global investment failures and transformations in modern biofuels; (iv) a socio-political, techno-economic, legal, and environmental analysis of biofuel energy industry in Europe; and (v) the role of participation in the responsible innovation framework for biofuels projects.

2.1. Assessment of Modern Bioenergy: Conditions and Synergies

Bioenergy has crucial advantages over other energy sources as a tool for poverty reduction in the least developed countries (LDCs). Though biomass resources are widely available in rural areas of LDCs, little seems to have been done with regards to the national modern bioenergy sector. Khatiwada et al. [8] assess the surplus agricultural residues availability for bioelectricity in six LDCs in Asia and Africa, namely Bangladesh, Lao People's Democratic Republic (Lao-PDR), and Nepal in Asia and Ethiopia, Malawi, and Zambia in Africa. The results indicate that the surplus biomass feedstock available from the agriculture sector could provide the total current electricity demand in Malawi alone, followed by Nepal (45%), Bangladesh (29%), Lao-PDR (29%), Ethiopia (27%), and Zambia (13%). Findings from the study show that providing access to sustainable energy in the LDCs to meet the SDG7 by 2030 might be a challenge due to limited access to technology, infrastructure, and finance. Site-specific investigations on how much agricultural residue could be extracted in an environmentally benign manner for bioelectricity and increased investment in the bioenergy sector are key potential solutions in a myriad of options required to harness the full energy potential in the LDCs. Provision of infrastructure/technology, mobilization of actors/stakeholders, regulatory action by governments, and market creation are essential for modernizing traditional bioenergy [9]. Conversion efficiencies, investment costs, and government policies (taxes and subsidies) are the key factors influencing the technological choice [10].

2.2. Water–Energy Nexus: Trade-Offs and Synergies

Land, water, and energy resource systems are integrated when it comes to the production of energy in general and bioenergy in particular [11]. In African nations, national and regional development targets for water and energy sectors seldom consider the nexus between the two, risking imbalances and inefficiencies in resource allocation and utilization. A typical example is the development and expansion of biofuel in the Wami Ruvu River Basin, Tanzania (WRB). Miraji et al. [12] applied the Water Evaluation and Planning (WEAP) model to the WRB to investigate the water–energy nexus (WEN), specifically, whether the development plan calling for biofuel expansion is a sound approach. The results indicate that WEN is much stronger in biofuel irrigation consuming 69.3% and 61% of total biofuel's water and energy requirements, respectively. By 2035, the nexus continues to be stronger, consuming 54.5% and 49% of total biofuel's water and energy requirements, respectively, and thus first-generation biofuels use much more resources in the growth than the refining process. An additional 768.2 million cubic meters of water and 413.4 million kWh of energy are needed for planned biofuel expansion, and reallocating water to biofuels in water-scarce regions causes related problems for other sectors, such as increasing water use for the industry, agriculture, and energy sector by 67%, 45%, and 9%, respectively, which could further exacerbate stresses on water and energy supplies in the basin. Biofuel generation relies heavily on energy imports, as it consumes substantially more energy than it produces. Policies should promote the coordinated development of sustainable biofuel programs that are less water-intensive with very low inputs of fossil fuels.

2.3. *Investments in Modern Biofuels: Failures and Transformations*

Substantial investments are needed in the promotion of modern biofuels [13]. There has been a heightened global interest in large-scale *Jatropha* cultivation for the past few decades, and this has encouraged investment in the crop in many developing countries [14]. *Jatropha* could be an alternative to conventional fuel as it contributes to climate change mitigation. However, *Jatropha* investment plans have not met global expectations. Antwi-Bediako et al. [15] review and synthesize the transformations and complexities in failed *Jatropha* spaces in the six major *Jatropha* investment destinations across the world—Mexico, India, China, Ethiopia, Mozambique, and Ghana. The review shows that the intended goal of establishing global *Jatropha* investments to serve as an alternative source of fuel failed because of the unexpected complexities of the hype. In all of the countries studied, promoters of *Jatropha* investments, including central government and private investors, subscribed to a "wait-and-see" approach for global *Jatropha* investments. Failure of the investments along with unmet expectations led to land-use changes from *Jatropha* to the cultivation of other crops (often food crops) or total land abandonment. The authors emphasize the importance of paying considerable attention to other feedstocks that might have a better future as alternatives to fossil-based energy for the deployment of sustainable bioenergy. Furthermore, the findings provide meaningful justification for policy- and decision-makers in the development space to tacitly reflect and appraise new investment initiatives or interventions before endorsement.

2.4. *Promoting Biofuel Industry: Interdisciplinary PESTLE Analysis Approach*

Biofuel production is expected to be an intrinsic confluence with the renewable energy sector in the coming years under the European regulations for renewable energy. Key standpoints for the promotion of biofuels are the reduction of national carbon emissions and rural deployment. Despite the jubilant outlook on biofuels for sustainable development, research efforts still tend to link the biofuel industry and regional growth. To overcome a variety of caveats in the development of modern bioenergy/biofuels, a novel approach, the "Political, Economic, Social, Technological, Environmental and Legal (PESTEL)" framework, has been used [16]. Achinas et al. [17] explore and review the biofuels industry through a PESTLE approach and discuss the interrelation between technological facets and sustainable deployment. The variety of the production characteristics and sustainability factors that were analyzed as a result of this approach highlight the complexity of the industry faced by policymakers and production companies. Socioecological aspects (origin of the feedstock, i.e., from agricultural land or a waste stream; food security; water scarcity; biodiversity loss), environmental governance (lifecycle emissions and land-use change), and economic landscape (yield and efficiency, cost of the conversion process) are found to be the key correlated dimensions in the assessment framework.

2.5. *Innovation in Biofuel Projects—A Participatory Framework*

The growth in biofuel investments brings with it concerns about the social and environmental impacts of the sector. Several tools and frameworks have been used to address these concerns, including the Responsible Research and Innovation (RRI) framework. Postal et al. [18] analyze whether this framework can be applied in contexts where local culture and values differently shape the freedom of speech and engagement, such as in developing countries in which biofuel innovation projects are often implemented. A literature review focusing on eight case studies of other authors is used to explore the role of "participation" as a structural element of the RRI framework and the impact of its absence where effective participation in the innovation development process is not possible. The authors describe the case of biofuel development, articulating four integrated dimensions of responsible innovation: anticipation, reflexivity, inclusion, and responsiveness. They highlight how the RRI inspirational normative framework, designed to influence innovation, is misused to judge its impacts. More than that, the conclusions of such misused applications reflect more the difficulties involved in applying guidelines than the responsible character of the innovation, whose impacts are usually defined upfront materially and measurably.

In conclusion, these five articles are a clear illustration of the challenges involved, and possible solutions, in the drive towards using modern bioenergy and strategies for sustainable development. Modern bioenergy is essential when the world is transforming towards environmentally friendly energy systems and low-carbon development. They should be developed while creating synergies with other sectors of the economy such as agriculture, industry, and rural development. The coordinated development and management of land, water, energy, and related resources in order to maximize economic and social welfare in an equitable manner without compromising the sustainability of vital ecosystems are critically important. Synergies and trade-offs should be balanced considering the optimal utilization of the resources for sustaining living standards and preserving the environment. The sustainability of the biofuel industry should be explored through a socio-political, techno-economic, legal, and environmental analysis approach, and the interactions between technological facets and sustainable deployment should be discussed. When it comes to large-scale biofuel investment, challenges need to be tackled while considering multilevel perspectives and global value chain frameworks. Finally, responsible and participatory innovation provides democratic governance of emerging biofuel technologies.

Author Contributions: Conceptualization, investigation, resources, and writing—original draft preparation: D.K. and P.P.; validation and writing—review and editing: D.K. and P.P. All authors have read and agreed to the published version of the manuscript.

Funding: This research received no external funding.

Acknowledgments: The editors express their gratitude to the participating authors and reviewers for their support and critical and constructive comments. This has significantly improved the quality of the collection as well as the academic output. The editors graciously extend their thanks to the editorial assistance office of MDPI for their support throughout the review and publication process of this Special Issue.

Conflicts of Interest: The authors declare no conflict of interest.

References

1. Reid, W.V.; Ali, M.K.; Field, C.B. The future of bioenergy. *Glob. Chang. Biol.* **2020**, *26*, 274–286. [CrossRef] [PubMed]
2. Souza, G.M.; Ballester, M.V.R.; de Brito Cruz, C.H.; Chum, H.; Dale, B.; Dale, V.H.; Fernandes, E.C.M.; Foust, T.; Karp, A.; Lynd, L.; et al. The role of bioenergy in a climate-changing world. *Environ. Dev.* **2017**, *23*, 57–64. [CrossRef]
3. Popp, J.; Lakner, Z.; Harangi-Rákos, M.; Fári, M. The effect of bioenergy expansion: Food, energy, and environment. *Renew. Sustain. Energy Rev.* **2014**, *32*, 559–578. [CrossRef]
4. REN21. REN21 Global Status Report—Renewables. 2020. Available online: https://www.ren21.net/gsr-2020/ (accessed on 1 January 2021).
5. IEA. *2020. World Energy Outlook 2020*; International Energy Agency (IEA): Paris, France, 2020; Available online: https://www.iea.org/reports/world-energy-outlook-2020 (accessed on 25 December 2020).
6. Masson-Delmotte, V.; Zhai, P.; Pörtner, H.O.; Roberts, D.; Skea, J.; Shukla, P.R.; Pirani, A.; Moufouma-Okia, W.; Péan, C.; Pidcock, R.; et al. Global warming of 1.5°C An IPCC Special Report on the impacts of global warming

of 1.5°C above pre-industrial levels and related global greenhouse gas emission pathways. In *The Context of Strengthening the Global Response to the Threat of Climate Change, Sustainable Development, and Efforts to Eradicate Poverty*; Intergovernmental Panel on Climate Change (IPCC): Geneva, Switzerland, 2018.
7. Faaij, A. Modern biomass conversion technologies. *Mitig. Adapt. Strateg. Glob. Chang.* **2006**, *11*, 343–375. [CrossRef]
8. Khatiwada, D.; Purohit, P.; Ackom, E.K. Mapping Bioenergy Supply and Demand in Selected Least Developed Countries (LDCs): Exploratory Assessment of Modern Bioenergy's Contribution to SDG7. *Sustainability* **2019**, *11*, 7091. [CrossRef]
9. Silveira, S.; Johnson, F.X. Navigating the transition to sustainable bioenergy in Sweden and Brazil: Lessons learned in a European and International context. *Energy Res. Soc. Sci.* **2016**, *13*, 180–193. [CrossRef]
10. Khatiwada, D.; Leduc, S.; Silveira, S.; McCallum, I. Optimizing ethanol and bioelectricity production in sugarcane biorefineries in Brazil. *Renew. Energy* **2016**, *85*, 371–386. [CrossRef]
11. Howells, M.; Hermann, S.; Welsch, M.; Bazilian, M.; Segerström, R.; Alfstad, T.; Gielen, D.; Rogner, H.; Fischer, G.; van Velthuizen, H.; et al. Integrated analysis of climate change, land-use, energy and water strategies. *Nat. Clim. Chang.* **2013**, *3*, 621–626. [CrossRef]
12. Miraji, M.; Li, X.; Liu, J.; Zheng, C. Evaluation of Water and Energy Nexus in Wami Ruvu River Basin, Tanzania. *Sustainability* **2019**, *11*, 3109. [CrossRef]
13. Nygaard, I.; Bolwig, S. The rise and fall of foreign private investment in the jatropha biofuel value chain in Ghana. *Environ. Sci. Policy* **2018**, *84*, 224–234. [CrossRef]
14. Purohit, P.; Dhar, S. *Biofuel Roadmap for India*; UNEP DTU Partnership: Copenhagen, Denmark, 2015.
15. Antwi-Bediako, R.; Otsuki, K.; Zoomers, A.; Amsalu, A. Global Investment Failures and Transformations: A Review of Hyped Jatropha Spaces. *Sustainability* **2019**, *11*, 3371. [CrossRef]
16. Song, J.; Sun, Y.; Jin, L. PESTEL analysis of the development of the waste-to-energy incineration industry in China. *Renew. Sustain. Energy Rev.* **2017**, *80*, 276–289. [CrossRef]
17. Achinas, S.; Horjus, J.; Achinas, V.; Euverink, G.J.W. A PESTLE Analysis of Biofuels Energy Industry in Europe. *Sustainability* **2019**, *11*, 5981. [CrossRef]
18. Postal, A.M.; Benatti, G.; Palmeros Parada, M.; Asveld, L.; Osseweijer, P.; Da Silveira, J.M.F.J. The Role of Participation in the Responsible Innovation Framework for Biofuels Projects: Can It Be Assessed? *Sustainability* **2020**, *12*, 10581. [CrossRef]

sustainability

MDPI

Article

Mapping Bioenergy Supply and Demand in Selected Least Developed Countries (LDCs): Exploratory Assessment of Modern Bioenergy's Contribution to SDG7

Dilip Khatiwada [1,*], Pallav Purohit [2] and Emmanuel Kofi Ackom [3]

1 Division of Energy Systems, Department of Energy Technology, School of Industrial Engineering and Management, KTH Royal Institute of Technology, Brinellvägen 68, 100 44 Stockholm, Sweden
2 Air Quality and Greenhouse Gases (AIR) Program, International Institute for Applied Systems Analysis (IIASA), Schlossplatz 1, A-2361 Laxenburg, Austria; purohit@iiasa.ac.at
3 UNEP DTU Partnership, UN City Campus, Department of Technology, Management and Economics, Denmark Technical University (DTU), Marmorvej 51, 2100 Copenhagen, Denmark; emac@dtu.dk
* Correspondence: dilip.khatiwada@energy.kth.se; Tel.: +46-8790-7464

Received: 23 October 2019; Accepted: 7 December 2019; Published: 11 December 2019

Abstract: Bioenergy can play an important role in achieving the agreed United Nations Sustainable Development Goals (SDGs) and implementing the Paris Agreement on Climate Change, thereby advancing climate goals, food security, better land use, and sustainable energy for all. In this study, we assess the surplus agricultural residues availability for bioelectricity in six least developed countries (LDCs) in Asia and Africa, namely Bangladesh, Lao-PDR, and Nepal in Asia; and Ethiopia, Malawi, and Zambia in Africa, respectively. The surplus agricultural residues have been estimated using residue-to-product ratio (RPR), agricultural residues lost in the collection, transportation and storage, and their alternative applications. We use a linear regression model to project the economic potential of bioelectricity. The contribution of bioelectricity for meeting the LDCs' electricity requirements is estimated in a time frame between 2017 and 2030. Our results reveal that the surplus biomass feedstock available from the agriculture sector could provide the total current electricity demand in Malawi alone, followed by Nepal (45%), Bangladesh (29%), Lao People's Democratic Republic (Lao-PDR) (29%), Ethiopia (27%), and Zambia (13%). This study also explores the complementarity and synergies of bioelectricity, SDG7, and their interlinkages with other SDGs. Findings from the study show that providing access to sustainable energy in the LDCs to meet the SDG7 by 2030 might be a challenge due to limited access to technology, infrastructure, and finance. Site-specific investigations on how much agricultural residues could be extracted in an environmentally benign manner for bioelectricity and increased investment in the bioenergy sector are key potential solutions in a myriad of options required to harness the full energy potential in the LDCs.

Keywords: energy access; least developed countries (LDCs); sustainable development goals (SDGs); agricultural residues; bioelectricity

1. Introduction

Access to modern and reliable energy sources is a prerequisite for improving living standards and promoting economic development. In 2015, the United Nations (UN) adopted the 2030 Agenda of Sustainable Development, which includes a set of 17 Sustainable Development Goals (SDGs). SDG7 aims at securing modern, affordable, and sustainable energy for all, thereby increasing the share of renewable energy (RE) in the global energy mix [1]. One in seven people still lacks electricity, and most of them live in rural areas of the developing world [2]. Energy is the main contributor (60%)

of the global greenhouse gas (GHG) emissions (climate change impact), and more than 40% of the world's population rely on unhealthy and polluting fuels for cooking [2]. These are mainly low-income countries that do not have access to modern energy where the majority of the population lives in rural areas.

Biomass (fuelwood, agricultural residues, and dung) is the main traditional energy source in the least developed countries (LDCs) and it is utilized inefficiently for cooking and heating purposes [3]. The LDCs are characterized by low income (Gross National Income (GNI) per capita), low level of human capital or human assets (health and education), and they are exposed to economic vulnerability (e.g., high population growth, geographic remoteness, natural disasters, etc.) [4]. The LDCs have a less diversified energy mix with traditional biomass accounting for 59% of the total primary energy supply (TPES) [3]. Despite the large share of traditional biomass in the primary energy matrix, the majority is combusted inefficiently. This, coupled with population growth, results in increased demands for more biomass (especially from the forest), thereby leading to deforestation. Additionally, the use of traditional biomass for cooking, especially in poorly ventilated facilities, results in indoor air pollution, lung diseases, injuries, and in severe cases, even death [5–7]. The bloc of LDCs also import a significant amount of petroleum (oil) products [8–10].

In this article, we assess the modern bioenergy potential for electrification, obtained from agricultural residues in the selected LDCs in Asia and Sub-Saharan Africa (SSA). Out of 47 LDCs globally, three representative countries in Asia: Bangladesh, Lao People's Democratic Republic (Lao-PDR), and Nepal; and three from SSA: Ethiopia, Malawi, and Zambia, are considered. Except for Bangladesh, the rest of the countries are land-locked (see Figure S1). Bangladesh has the highest population density (i.e., 1115 capita per square km), while Zambia has the least population density, with only 23 capita per square km. The majority of people live in rural areas and agriculture is one of the key contributors to the Gross Domestic Product (GDP) in their respective countries. Ethiopia, Lao-PDR, and Bangladesh have more than 6% GDP growth rate during the past decade [11], while the economic growth rates of Nepal, Malawi, and Zambia are not stable [12] (Figure S2).

Low-income countries are typically agrarian economies and the agriculture sector offers the highest employment rate, for example, 72% in Malawi, 70% in Nepal, and 40% in Bangladesh [13]. They produce plenty of agricultural commodities/crops. As a result, there are agricultural residues in farms and co-products in agro-industries that could be utilized for energy generation. Per capita electricity consumption of the selected LDCs is far below the world average, i.e., 3150 kWh in 2017. For example, Zambia has the highest per capita electricity consumption (i.e., 730 kWh/capita) while Ethiopia has the lowest (i.e., 90 kWh/capita) among the selected LDCs in 2017 [10].

However, relatively few studies have explored bioenergy potential comparatively in a cluster of the selected LDCs in Asia and Africa. Huda et al. [14] have presented the prospects and technologies related to the biomass energy in Bangladesh. Halder et al. [15] assessed the biomass energy resources and related technologies and practices in Bangladesh. Toth et al. [16] investigated the use of agroforestry and fuelwood in Malawi. Shane et al. [17] assessed the bioenergy potential from biomass in Zambia. In Nepal, Gurung and Oh [18] reviewed the conversion of traditional biomass into modern bioenergy systems (improved cooking stoves and biogas). Khatiwada et al. [19] evaluated power generation from sugarcane biomass in Nepal. Ackom et al. [20] assessed the biomass resource potential in Cameroon from sustainably extracted agricultural and forest residues. Sasaki et al. [21] estimated the woody biomass and bioenergy potentials in Southeast Asia, including Lao-PDR. The Least Developed Countries Report 2017 presents the energy status of LDCs [3]. However, an exploratory assessment of modern bioenergy's contribution to SDG7, considering agricultural biomass/residues in the LDCs, has not been done yet.

The primary objective of this study is to assess agricultural residues availability and their potential for bioelectricity in the studied LDC countries in Africa and Asia. The contribution of bioelectricity for meeting SDG7 in their respective countries is analyzed. Bioelectricity potential from the major crops such as rice, maize, wheat, sugarcane, cassava, and potatoes is investigated. The paper investigates four interrelated questions: (i) How much is the supply of agricultural residues for energy production (i.e., mapping bioenergy supply and demand) in the selected LDCs?; (ii) How can bioelectricity help achieve SDG7?; (iii) How can bioelectricity be developed in LDCs in synergies with other SDGs?; and (iv) how can we create an enabling environment for the sustainable deployment of bioelectricity derived from environmentally benign agricultural residues in the selected LDCs? The contrasting cases in terms of diverse topography, population/demography, natural resources, development plans, and national priorities provide knowledge contribution for the studied countries and possibly other nations with similar biomass resources and circumstances. This study also provides important insights on the untapped agricultural biomass and informs relevant stakeholders (agro-industries, development partners, and policymakers) in realizing the full potential of renewable electricity in synergy with agricultural and rural development, climate change mitigation, etc. This paper is structured as follows: first Section 1, the introduction, provides the background and sets the rationale for the study. In Section 2, we contextualize the access to clean and modern energy services. The need for modernizing bioenergy and the status of SDG7 are also presented in Section 2. Section 3 assesses the sources of electricity generation and future projection of electricity demand up to 2030. Section 4 provides a framework for estimating the gross and net agricultural residue availability for bioelectricity. The technical and economic potential of agricultural residues for biomass power generation is also presented in Section 4. Section 5 investigates the bioelectricity potential in the selected LDCs under different scenarios towards meeting SDG7. It explores the complementarity and synergism of bioelectricity, SDG7 and their interlinkages with other SDGs. Finally, concluding remarks are made in order to find a way forward to achieve the SDG7 in the selected LDCs.

2. Energy and Development in the LDCs

2.1. Economy and Energy Profile

Energy plays a key role in the economic development of countries. The energy transition from traditional biomass to modern bioenergy is essential for industrial and economic growth [22,23]. The majority of LDCs do not have access to modern energy services and they are characterized by relatively low Human Development Index (HDI). The HDI—measure of health, education, and standard of living (GNI per capita)—is quite less for the selected LDCs, ranging from low human development countries, such as Ethiopia (0.46) and Malawi (0.48), to medium human development countries, such as Nepal (0.57), Bangladesh (0.61), Lao-PDR (0.60), and Zambia (0.59). The population has been continuously increasing in the selected LDCs, with average annual growth rates ranging from 1.0% (Nepal) to 2.9% (Zambia) during the past two decades (1998–2017); refer to Table 1. In spite of the rapid urbanization, the majority of the population lives in rural areas (Table 2). Energy consumption and GNI in the LDCs are low compared to developing and industrialized countries [11]. Table 1 provides the socio-economic metrics, energy, and related emissions of the six LDCs in Africa and Asia. Refer to Figure S3 which depicts the historic trend of urbanization and population growth in the selected LDCs.

Bangladesh has an average annual GDP growth rate of 7.8% since 1990 and its GDP per capita was US$1564 in 2017 [11]. Lao-PDR has remarkably achieved an annual economic growth rate of around 8% during the period 2000–2016 and its GDP per capita was US$2424.5 in 2017 [9,11]. Ethiopia has also made substantial progress in the economic development in the last decade with a 4-fold increase in GDP per capita [11]. However, the country still has a low GDP per capita (US$768 in 2017) [11]. Malawi has the lowest GDP per capita (US$356.5) even though there is an annual average growth of 3.1% in the last decade [11], refer to Figure S3.

Table 1. Socio-economic metrics, energy, and related emissions in the selected least developed countries (LDCs).

Parameter	Unit	Bangladesh	Lao-PDR	Nepal	Ethiopia	Malawi	Zambia	Reference Year(s)	Source
		Value							
Population	Million	159.7	7.0	27.6	106.4	17.7	16.9	2017	[11]
Average population growth rate (1998–2017), per year	%	1.3	1.6	1.0	2.8	2.8	2.9	1998–2017	[11]
Gross domestic product (GDP)	Billion US$	249.7	16.9	24.9	81.7	6.3	25.9	2017	[11]
GDP/population (current US$)	GDP/capita	1564	2425	901.	768	357	1535	2017	[11]
Real GDP growth, (2009–2016, selected year)	% (average 2009–2016)	6.2	7.7	3.9	10.2	4.7	6.1	2009–2016	[24]
	%	7.2	7.3	4.6	7.3	4.4	4.2	2017	
Foreign direct investment inflows	Millions of US$	2152	813	198	3586	277	1091	2017	[24]
Poverty rate (population below US$1.90 a day)	% of the total population	15	23	15	27	71	58	2018	[24]
Employment in the agricultural sector	% of total employment	39	61	72	68	85	53	2017	[24]
Human development index	–	0.61	0.60	0.58	0.46	0.48	0.59	2017	[25]
Total primary energy supply (TPES)	Mtoe	39.5	4.8	12.6	41.0	4.1	11.1	2016	[10]
Energy consumption/population	GJ/capita	8	57	6	3	2	11	2017	[26]
Net energy imports	Mtoe	6.4	N/A *	2.9	4.1	N/A *	1.2	2016	[10]
CO_2 emissions from fossil energy	$MtCO_2$	88.0	2.0	9.0	13.0	1.4	4.7	2017	[27]
CO_2 emissions/population	t CO_2/capita	0.48	0.30	0.30	0.12	0.08	0.29	2014	[11]

* N/A: No data is available.

Table 2. Share of Gross Domestic Product (GDP) in the economy, population, employment, and land covered by the agriculture sector.

Parameters	Bangladesh	Lao-PDR	Nepal	Ethiopia	Malawi	Zambia	Reference Year	Source
			Share (%) of GDP in the economy					
Agriculture, forestry, and fishing	13.1	15.7	25.0	31.1	26.1	2.6		
Industry	28.5	31.5	13.4	27.3	14.4	36.3	2017	[11]
Services	53.0	41.6	50.3	36.5	52.4	54.1		
Others	5.4	11.2	11.3	5.1	7.1	7.0		
			Agriculture sector					
% of total employment	39.1	61.3	71.7	68.2	84.7	53.3	2017	[24]
% of land area cover (2016)	70.6	10.3	28.7	36.3	61.4	32.1	2016	[11]
% of the total population in the rural area	64.1	65.6	80.7	79.7	83.3	57.0	2017	[11]

Due to economic progress, urbanization, and rising population, energy supply and consumption have also increased in the LDCs. In Bangladesh, energy consumption increased by around 50% in the last decade (2006–2016) [10]. Nepal's total primary energy supply increased by almost 40%, escalating the share of commercial energy (hydro and fossil oil) from 9.7% to 18.5% [10]. In Zambia, the TPES has increased by 3.6% annually from 7.78 million tonne oil equivalent (Mtoe) to 11.08 Mtoe between 2006 and 2016, whereas total final energy consumption (TFEC) has also grown by 3.7% in the last decade [10]. In Ethiopia, in spite of the continuous increase in the energy supply (3.5% annual increase in the last 10 years), TPES per capita has almost remained the same since 1990, with a value of 0.5 tonne oil equivalent (toe) per capita [10].

In Bangladesh, of the total TPES (i.e., 39.54 Mtoe), natural gas has the largest share with a contribution of 58.4%, followed by biomass/biofuels (23.9%), petroleum oil (13.3%), and coal (4.3%) [10]. Out of the total 5.58 Mtoe, coal, biomass, and hydroelectricity are the major sources of the TPES in Lao-PDR, with the corresponding shares of 32.3%, 29.0%, and 22.1%, respectively [9]. In Nepal, biomass, imported fossil (coal and oil products), and electricity are its main sources of the TPES (12.65 Mtoe in 2016). Commercial energy amounts to 18.5% of the total in the form of fossil fuels (15.6%) and electricity (2.9%) in the country [10]. In the LDCs, net energy imports (petroleum products) has drastically increased in the last decade [10], thereby spending a huge amount of foreign currency on the import of petroleum products. The import of petroleum products has increased from 1.75 Mtoe to 23.99 Mtoe in one decade (between 2006 and 2016) in Ethiopia [10]. Domestic (or residential) sector has the highest share in the TFEC, e.g., 92.7% in Ethiopia, 82.7% in Nepal, 65% in Zambia, and 58.5% in both Bangladesh and Lao-PDR. Energy is mainly used for cooking and lighting purposes. On the other hand, transport and industrial sectors have relatively low shares in the TFEC: Bangladesh (transport, 12.5%; industry, 28.9%), Nepal (transport, 10.1%; industry, 7.2%), Ethiopia (transport, 4.1%; industry, 3.2%), and Zambia (transport, 4.6%; industry, 30.4%) [10].

Agriculture is the major source of economic activity in the LDCs. Over 40% of the population in the selected LDCs is employed in this sector [11]. Table 2 shows the share of GDP in the economy, population employed in the agricultural sector, and rural population and land area in the LDCs. Around 80% of population live in rural areas in Nepal (80.7%), Ethiopia (79.7%), and Malawi (83.3%), whereas other LDCs also have the majority of the rural population: Bangladesh (64.1%), Lao-PDR (65.6%), and Zambia (57.0%). The agricultural sector provides up to 85% of employment in the LDCs (Table 2). In Nepal, the agricultural sector accounts for one-fourth of GDP's contribution, whereas the industrial and service sector has 13.5% and 50.3%, respectively. Regardless of the countries' large population employed in the sector, the contribution to the national GDP is relatively low. In Zambia, the agricultural sector merely contributed to the economy with a small share of 2.6%, while the population employed in the sector was 53.3%. This is mainly due to the lack of modernization, productivity (production/hectare), and efficiency of the agricultural sector. There are still many people in the LDCs who are living under the national poverty lines [11]. For example, poverty rate (population below US$1.90 a day) is 14.8% (Nepal), 26.7% (Ethiopia), and 71.4% (Malawi) (see Table 1).

2.2. *Biomass in the Primary and Final Energy Consumption*

As mentioned earlier, rural dwellers heavily rely on traditional fuels for cooking. Biomass and wastes are the key primary energy sources in the selected LDCs, notably Nepal, Ethiopia, Malawi, and Zambia. The shares (in the TPES) of biomass/biofuels were 23.5% (Bangladesh), 34.0% (Lao-PDR), 73.7% (Nepal), 87.7% (Ethiopia), 88.2% (Malawi), and 75.4% (Zambia). Refer to Table S1 for the TPES in the selected LDCs by fuel type.

In Lao-PDR, the use of biomass has drastically decreased from 78% to 34% between 2000 and 2015 [9], which is mainly due to the rapid expansion of hydropower plants and increased consumption of fossil fuels. In Nepal, traditional biomass has the largest share of the TFEC, amounting to 77.6%, followed by oil products (12.5%), coal (4%), and electricity (3.4%) in 2015 [28]. Figure 1 presents the share of fuel types in the total energy matrix in Nepal. Approximately 70% of the population use fuelwood for energy consumption, mainly for cooking [28]. These numbers denote the present low level of industrial and economic activities in the country.

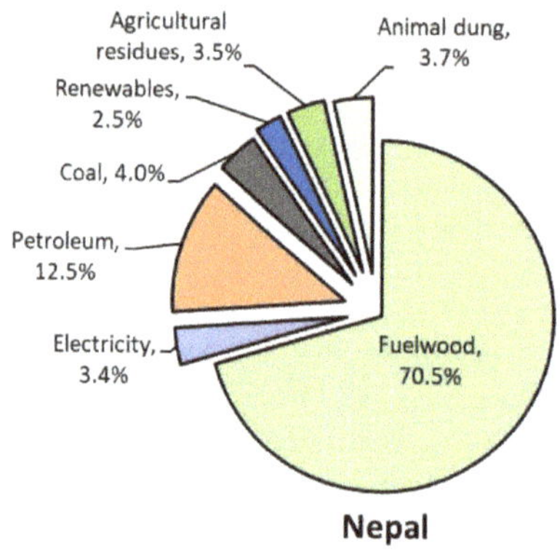

Figure 1. Distribution of energy consumption by fuel type (in 2015) in Nepal [28]. (Note: Total energy consumption was 500 PJ).

In Ethiopia, the share of biomass and biofuels was 87.5% in the TFEC (by source) in 2017, followed by oil products (9.5%) [10]. It should be noted that Ethiopia has only 12.5% of forest land while 36.3% area was covered by agricultural land [11]. In Malawi and Zambia, traditional biomass remains the dominant source of primary energy (Figure 2). Zambia has a share of 75.4% of the traditional biomass in the TPES [10], whereas Malawi has a gigantic share of 88.22% [8]. It is noteworthy to mention that the share of biomass in the TPES is almost the same in the last ten years in Malawi and Zambia. Among the total TFEC, traditional biomass has a significant role in the LDCs (see Figure 3). Table S1 summarizes the share of the TPES in the selected LDCs by fuel source.

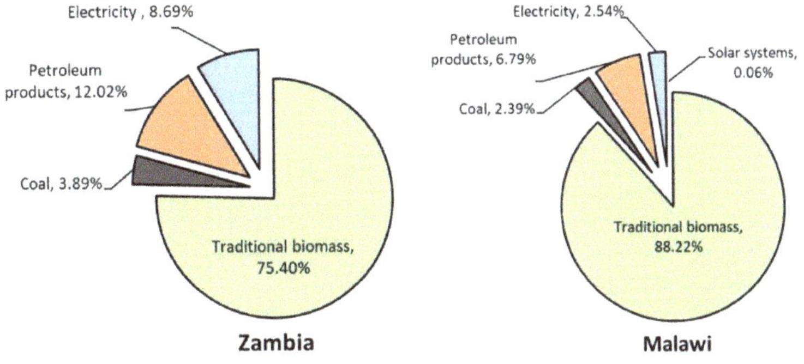

(a) (b)

Figure 2. Total primary energy supply (TPES) by fuel types in (**a**) Zambia and (**b**) Malawi [8,10].

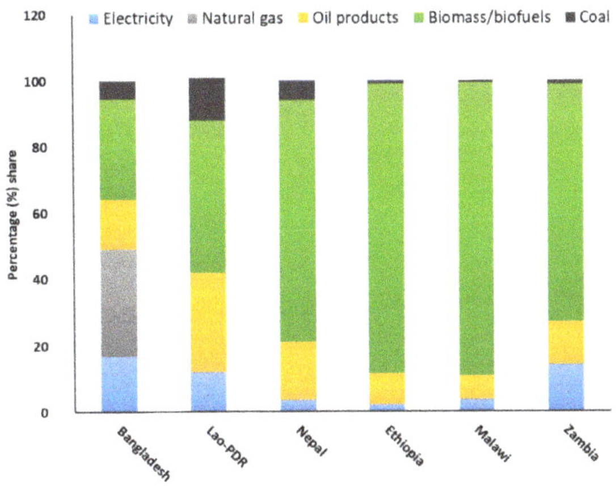

Figure 3. Total final energy consumption (TFEC) by fuel type in 2017 (for Bangladesh, Nepal, Ethiopia, and Zambia) [10], and in 2015 for Lao-PDR [9] and Malawi [8,29].

2.3. The Need for Modernizing Traditional Biomass Utilization

As mentioned in the previous section, the majority of people live in rural areas in the LDCs using traditional biomass fuel as a major source of energy. In the SSA, the share of biomass in the household energy consumption is around 90% [30]. The incomplete burning of biomass poses a severe health impact due to the release of small smoke particles and carbon monoxide (CO) [5,6]. Biomass burning is a significant air pollution source, with global, regional, and local impacts on air quality, public health, and climate [31]. Therefore, it is important to avoid an inefficient burning of biomass for cooking. Furthermore, the open burning of biomass (agricultural residues) in the field also causes air pollution, public health risk, and climate impacts [32,33]. There are also problems related to deforestation and land degradation while using excessive fuelwood for cooking.

The expansion of agricultural land and the high demand for fuelwood as fuel promote deforestation. Figure 4 shows the trend of agriculture and forestland in the selected LDCs. The farmland is increasing in all countries, and consequently, the forest cover is decreasing, except in Lao-PDR. Thus, traditional biomass shall have a considerable impact on the environment (i.e., deforestation and desertification) and public health (i.e., indoor air pollution). Modernization of bioenergy can add value to existing resources and serve to meet increasing energy demand, as well as create jobs and reduce poverty [34].

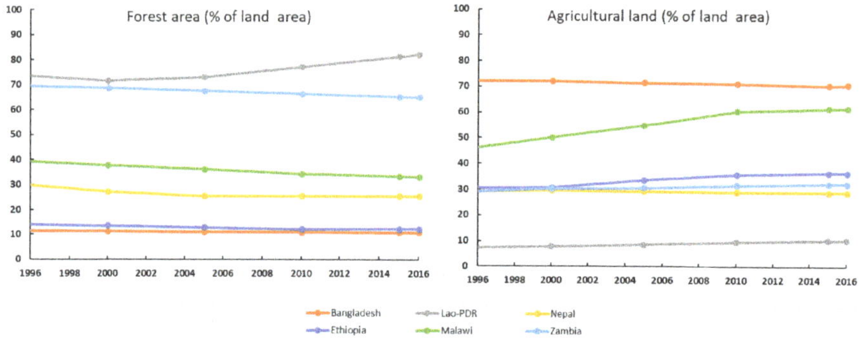

Figure 4. Land covered by agricultural and forest land (% of land area) [35], also refer to Table S2.

An improved cooking stove (ICS) can help improve burning efficiency and reduce health hazards, and it has a low rank in the energy ladder which describes the selection of fuel choices as per the variations in the economic status of the individual household [36,37]. A transition toward cleaner and more efficient forms of energy is achieved by moving up the energy ladder. Household incomes (economic well-being) and availability of cleaner fuels also play an important role in switching and/or diversifying energy services. We assume that there would be government incentives for promoting locally produced bioelectricity and rural households can afford to buy/utilize it. Bioelectricity from agricultural residues has the potential to contribute to resource efficacy, protection of forests, reduce GHG emissions, and protect human health or avoid indoor air pollution in rural areas of the LDCs.

2.4. Defining SDG7—State of Art and Perspectives and the Possible Role of Modern Bioenergy

The 2030 Agenda for Sustainable Development adopts 17 SDGs, which aim to end poverty, improve health and education quality, reduce inequality, enhance economic growth while mitigating climate change, and preserving ecosystem services, etc. Energy is a key enabler for several SDGs including economic development and social wellbeing. SDG7 (ensuring access to affordable, reliable, sustainable, and modern energy for all) has five targets and six indicators. The SDG7 targets and indicators are outlined in Table S3.

The SDG7 is strictly related to several other SDGs, notably climate change (SDG13), poverty eradication (SDG1), elimination of hunger (SDG2), gender equity (SDG5), health (SDG3), and clean water (SDG6). In this section, we present the progress of the selected LDCs in their efforts towards achieving SDG7—especially the targets 7.1 (7.1.1 and 7.1.2), 7.2, and 7.3. Targets 7.A and 7.B consider the means of implementation, mainly financing. Table 3 summarizes the current status of SDG7 in the selected LDCs.

Table 3. Status of United Nations Sustainable Development Goal 7 (SDG7) in the selected LDCs.

SDG7 (Targets/Indicators)	Bangladesh	Lao-PDR	Nepal	Ethiopia	Malawi	Zambia	Reference Year
SDG7.1.1 Access to electricity							
of the total population (%)	88	94	96	44	13	40	2017
of the urban population (%)	100	100	99	97	58	75	2017
of the rural population (%)	81	91	95	31	4	14	2017
SDG7.1.2 Access to clean fuels for cooking (% of the population)	19	6	29	3	2	16	2017
SDG7.2 RE (% of TFEC)	34	52	79	92	79	89	2016
SDG7.3 Energy efficiency (MJ per US$ PPP 2011)	3.1	5.9	8.1	13.1	4.2	7.7	2016

Source: [38], Also refer to Table S3, Table S4, Table S5, Table S6.

2.4.1. Access to Electricity

There is significant progress globally in the access to electrification [38]. The recent report by World Bank shows that the global electrification rate reached 89%, with a resultant drop in the number of people without electricity to 840 million compared to 1 billion in 2016 and 1.2 billion in 2010 [38]. It is projected that only 8% of the world population will not have access to electricity in 2030 and 90% of them will be in SSA [38]. Around 97% of the urban population in the world has access to electricity, while only 76% of the rural population had access to electricity in 2017 [38]. Figure S4 presents the historic progress in the global electricity access since 1990.

Historically, the LDCs are characterized by a lower rate of electrification, and access to electricity is limited compared to the global average. Figure S4 shows the evolution of electricity access in the selected LDCs, SSA, and the world. It depicts that there is a rapid growth in electricity access to households in the studied Asian countries: Bangladesh, Lao-PDR, and Nepal.

As depicted in Figure S4, countries in Asia have a higher rate of electricity access compared to the countries in SSA. In spite of the progress made in the last years, Ethiopia, Zambia, and Malawi still have relatively low electrification access rates of 44%, 40%, and 13%, respectively (see Table 3). There is a huge disparity in access to electricity. The urban population in Nepal, Bangladesh, and Lao-PDR have almost 100% access to electricity, whereas rural people in Malawi surprisingly have very limited electricity access. The majority of the population lives in rural areas where the rate of electricity in rural areas is low. Only 4% and 14% of the rural population of Malawi and Zambia, respectively, had access to electricity in 2017 [38]. It is worthwhile to mention that electricity is mainly used for lighting. Access to electricity for productive uses such as irrigation and other agricultural activities can also help promote rural development in the LDCs. Therefore, it is important to electrify the rural population for economic development and social well-being.

2.4.2. Access to Clean Fuels for Cooking

As mentioned in Section 2.2, traditional biomass is the main source for cooking and heating in the LDCs. The use of traditional biomass and inefficient burning of biomass is often associated with negative consequences such as deforestation, indoor air pollution, ill-health (lung-related diseases), injuries, and sometimes death (Section 2.3). Table 3 presents the status of access to clean fuels in the LDCs. On a global scale, approximately 3 billion of the population did not have access to clean fuels and technologies in 2017, and the majority of the population resided in the LDCs [38]. Malawi (2%), Ethiopia (3%), Lao-PDR (6%) had the lowest access to clean fuels and technologies for cooking in 2017. At the current annual rate of progress, it would be difficult to meet the SDG7.1.2 (target of universal access to clean cooking) [38]. Thus, cleaner fuels and efficient cooking technologies should be promoted with innovative scale-up schemes (models) for multiple benefits [39].

2.4.3. Renewable Energy Share in the Total Final Energy Consumption (TFEC)

All six LDCs considered in this study have more than the world average of around 20% share of renewables in TFEC. Four LDCs including Ethiopia, Zambia, Malawi, and Nepal had the largest share of RE which was more than 75% in 2017 [38], also refer to Table 3. Bangladesh had a 34% RE share in the TFEC, which is the lowest among the six LDCs considered in the analysis, whereas the RE share in Ethiopia (91.1%) was the highest, followed by Zambia (88.5%), Nepal (79.2%), and Malawi (78.5%).

2.4.4. Energy Efficiency (Energy Intensity)

Energy intensity is a proxy indicator to measure energy efficiency at national levels, which is the ratio of the TPES to the GDP, measured at purchasing power parity (PPP) in constant 2011 US$. This indicates how much energy is used to produce one unit of economic output, and hence a lower ratio depicts that less energy is required to produce one unit of economic product. LDCs have shown the fastest decline in energy intensity compared to industrialized and developed countries. Table S6 shows the trend in primary energy intensity in the LDCs. The increased share of industries and consumption of commercial (modern) energy production by fuel type in the national economy have contributed to making significant progress in the energy intensity.

3. Source of Electricity Generation, Consumption Pattern, and Future Demand Projection

In this section, we present the source of electricity generation and consumption patterns in the selected LDCs. The future projection of electricity generation and demand is also presented.

3.1. Electricity Generation and Consumption Pattern in the LDCs

The major source of electricity generation in the LDCs analyzed in this study is largely hydropower, except in Bangladesh (Figure 5). Nepal, Lao-PDR, Ethiopia, and Zambia have abundant hydroelectricity potential. Nepal and Zambia have almost 100% hydropower share in their electricity mix. However, electricity is mainly produced from natural gas in Bangladesh, in contrast to other LDCs, while Lao-PDR, Ethiopia, and Malawi have a small share of electricity derived from oil products and other renewables.

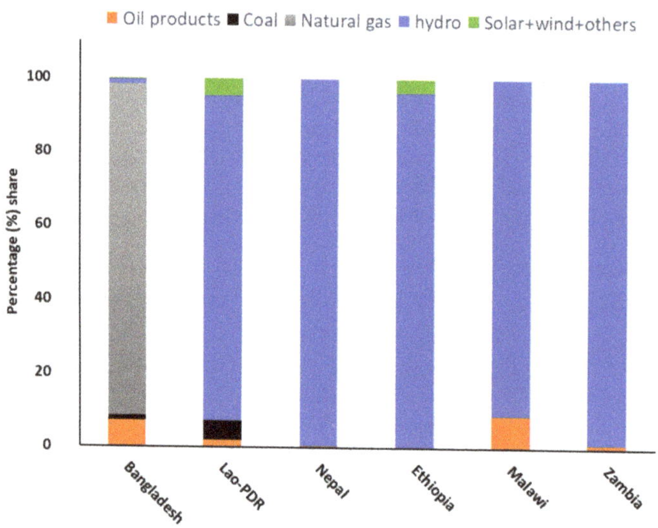

Figure 5. Composition of gross electricity production in the LDCs by fuel source, 2012–2014. Source: [24].

In Bangladesh, the installed capacity of power generation was 12,540 MW in 2016, including 600 MW energy export and renewables [40]. At present, the public sector has a share of 53% of the installed capacity and the private sector add up to another 42% of the installed capacity, whereas the remaining demand is met through the imports. The country had generated 64,327 GWh of electricity in 2016 [10]. Out of this total production, natural gas carried the majority share with 82%, followed by fossil oil (15%). Renewables (hydro, solar, and wind) had only around 1% of the total production. As regards the total sectoral retail electricity consumption, domestic sector held the major share with a 50.2%, followed by the industry (35.5%), commercial, (9.3%), and agricultural sectors (3.1%) [40]. In Lao-PDR, the total installed power generation capacity was 6418 MW in which hydropower had a share of 70.7% and combustible fuels had a share of 29.3% in 2016 [41]. Out of the total electricity production of 16,302 GWh by source, hydropower contributed 86% and the remaining came from coal power [9]. In Lao-PDR, a major portion of the electricity produced is exported to neighboring countries, mainly to Thailand. The export of electricity accounted for around 75% of the total electricity generated in the country. Electricity from coal has been used for export purposes since 2013. When it comes to the domestic consumption of electricity, 4248 GWh was consumed in several sectors: the majority of the consumption in industries (41.1%), followed by households (37.6%), and service (20.4%) [41]. It should be noted that the country also imports electricity during the dry season when there is less production of electricity from hydropower plants [9]. Nepal had 1132 MW installed capacity of electricity (including isolated/grid-connected or hydro/thermal) in 2016 [28,42,43]. Isolated energy systems (i.e., micro-hydro and solar home systems) serve remote rural areas and constitute 5.5% (i.e., 62.2 MW) of the total installed capacity. Out of the total annual electricity demand of 6258 GWh, Nepal domestically produced 4476 GWh, while 2582 GWh was imported from India in 2017 [43]. In the national grid, electricity from diesel thermal plants has a negligible share due to high operation and maintenance costs. The residential sector was responsible for the largest share of electricity consumption (45.2%), followed by the industrial sector (36.0%). Over the last years, electricity demand has significantly increased, including the peak demand, i.e., the highest power demand that occurs over a certain period [43]. It should be noted that Nepal used diesel generators to provide electricity in urban centers during load shedding periods, leading to huge local air pollution in cities [44].

In Ethiopia, the total installed capacity of electricity generation was 4238 MW in 2016 [41]. Hydropower has the largest share of around 89.9% [41]. The rest is from renewables (7.81%, mainly wind and geothermal) and diesel power plants (2.34%). The country consumed 8802 GWh of electricity in 2016 [41]. Electricity consumption is mainly in urban households and small industries. The household/domestic sector used almost two-third of electricity and the remaining amount was utilized by industrial/commercial sectors. In Malawi, the total installed capacity for large hydropower was 345.5 MW, whereas 23.8 MW was from bagasse and small hydro in 2014 [45]. Net production of electricity in 2016 was 2058 GWh and the household sector was the main consumer (878 GWh, 55.7%) followed by industries (699 GWh, 44.3%) [41]. Zambia had the total installed capacity of 2829 MW in 2016: 84.3% from hydropower and 15.7% from combustible fossil fuels with the generation of 11,695 GWh of electricity in 2016 [41]. Hydropower dominates the total electricity generation with a share of 94.3%. The remaining electricity comes from coal and oil. Around 60% of the electricity is used in industry and the remaining 40% in the household sector. The country also imports electricity from the neighboring country depending upon the availability of water in rivers. For example, the country imported 1391 GWh of electricity in 2016 [41]. Recently, Zambia has experienced an extended load-shedding (power-cuts) in the country, reducing its electricity consumption by 30% in 2015 [46]. This is mainly due to low water levels in reservoirs and declining water flows in rivers [47]. The total installed power generation capacity by fuel type in the selected LDCs is presented in Figure 6.

3.2. Future Projection of Electricity Generation and Demand in the LDCs

For meeting the universal access to cleaner and renewable energy in the LDCs, there should be an increased generation of electricity supply. We have taken into account the respective countries' plans and policies in order to project the electricity generation until 2030. Anticipated trends towards increased urbanization and industrialization, coupled with continued economic growth and population rise, would trigger the rise in electricity consumption in the LDCs. As the bloc of countries is also aiming to uplift their economies from the LDCs to medium-income countries, electricity is essential for this transition. To assess the electricity generation and peak load demand in the near future we have used the economic growth and electrification pattern of selected LDCs analyzed in this study. Figure 7 presents the electricity generation (TWh) and peak load demand (MW) in the selected LDCs until 2030, also refer to Table S7.

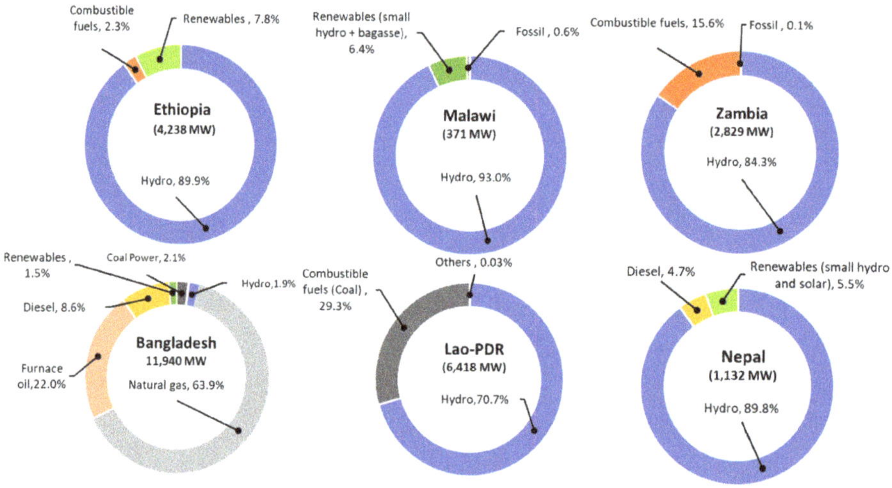

Figure 6. Installed power generation capacity by fuel type in the selected LDCs in 2016/2017; Authors' illustration based on data from the following cited sources for Ethiopia, Malawi, Zambia, and Lao-PDR: [41]; for Nepal: [28,43], and for Bangladesh: [48].

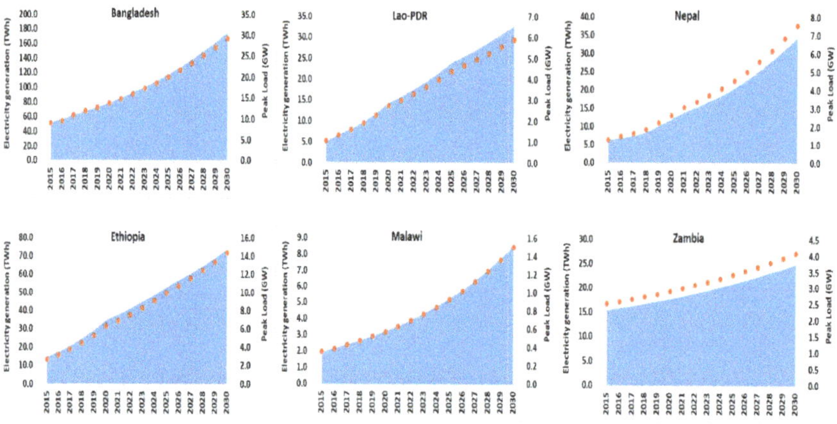

Figure 7. Projection of electricity generation (TWh, primary y-axis) and peak power (MW, secondary y-axis) in the selected LDCs (source: authors' compilation based on the results of projection). Also, Refer to Table S7.

Bangladesh has increased its electricity demand at an annual average of 8.26% since 2000 [40]. For the projection, we use an annual growth rate of 7.85% in peak load, which is also aligned with the current trend [48]. Energy production and peak power would be 175,926 GWh and 29,264 MW, respectively, in 2030 (see Figure 7). As mentioned above, Lao-DPR's net export of electricity was around 75% of the total production. Domestic electricity demand is also increasing—it has increased by a factor of thirteen from 2000 [9]. The increase is mainly in the household (domestic) and the industrial sector. According to the Lao-PDR's demand forecast (2016–2030), it is estimated that there would be a 13.4% annual increase in electricity generation in the country [49]. Electricity generation would be 32,923 GWh by 2030 and the corresponding peak load is set at 5892 MW. The export of electricity is not considered in the projection.

Nepal Electricity Authority (NEA), the state-owned public company, is mainly responsible for carrying out load forecasting, the generation planning of the power system of Nepal. Peak demand has doubled from 722 MW in 2008 to 1444 MW in 2017 [43]. However, the generation of hydroelectricity has not sufficiently met the country's increasing demand. The demand for electricity is expected to rapidly increase with an average annual growth rate of 12.48% due to the nation's economic and industrial development [43]. Peak load is estimated to be 7542 MW with the total electricity generation amounting to 34,355 GWh in Nepal by 2030 (see Figure 7). Ethiopian Electric Power Corporation has projected the national electricity demand in the form of energy generation (GWh) and peak load (MW) from 2012 until 2037 in different scenarios [50]. For example, in the base case, the peak demand forecast grows from 2657 MW in 2015 to 14,372 MW by 2030. This represents an average annual growth rate of 12.3% in energy generation (GWh) and 10.3% in peak load (MW).

There are a few old studies and reports which investigate the trend and future projection of the electricity demand and generation in Malawi. Taulo et al. [8] undertook both short and long-term energy demand forecasts for Malawi using the Model for Analysis of Energy Demand (MAED) for the period of 1999–2011 and 2008–2030. The grid electricity demand has increased from 1032 GWh to 1888 GWh from 1999 to 2011, representing an average annual growth rate of 6% [8]. As per the UN report, the net production of electricity has increased from 1793 GWh to 1875 GWh between 2011 and 2016 with the corresponding 288 MW (2011) and 352 MW (2016) generation capacity in Malawi [41]. The peak load in Malawi was 323 MW in 2016 [51]. Electricity Supply Corporation of Malawi (ESCOM), a state-owned authority, is responsible for the supply and distribution of electricity in Malawi. The government of Malawi in 2010 estimated that the projected electricity demand would be 598 MW in 2015, 874 MW in 2020, 1193 MW in 2025, and 1597 MW in 2030 [8]. However, the demand has been unmet, leading to load shedding up to 12 hours a day. In this study, the projected installed capacity (peak load) of electricity generation is estimated to 1500 MW in 2030 (Figure 7), which is approximately a 10% annual increase from the current capacity of 352 MW 2016. The corresponding 8541 GWh would be generated by 2030, considering a capacity factor of 65% [52]. Zambia developed a Power Systems Development Master Plan in 2011 [52]. Energy demand forecasted by the government of Zambia shows that there would be a 4.1% increase in the electricity production until 2030, reaching the generation up to 21,481 GWh. The projected peak load was 4066 MW (in 2030) which is around 2.5 times higher than that of 1600 MW in 2008. While performing the estimation of electricity demand (GWh), we consider the given peak loads and load factors from Power Systems Development Master Plan, which lies between 68% and 74% [52].

4. Estimating the Bioelectricity: Methodological Approach and Data Sources

When it comes to assessment of bioelectricity from agricultural residues, this study looks at three interrelated questions: (i) what is the supply of biofuels (agricultural residues) in the selected LDCs?; (ii) what would be technical potential of supply of bioelectricity?; and (iii) what is the economic potential of the bioelectricity in the selected LDCs considered in this study? The first research question is more related to the availability of agricultural residues. Nevertheless, both the technical and economic potential of bioelectricity require a detailed technical evaluation of surplus biomass feedstock availability for energy applications.

The conversion of agricultural residues to biomass feedstock for electricity generation is developing as a potential form of bioenergy. Biolectricity can be mainly produced through the combustion of lignocellulose feedstock which is obtained from biomass sources such as agricultural products and residues, plantation forests, sawmill residue, and native forests. In order to estimate the amount of agricultural residues that can be used for the production of bioelectricity, we use the methodology developed by Tripathi et al. [53] that was further improved by Purohit et al. [54] after taking into account the crop and its residue production, environmental constraints, and their competitive uses. The gross agricultural residue availability essentially depends upon the area under the crop, yield, and residue to product ratio for the crop. Therefore, the gross agricultural residue, GAR_i, availability for the *i*th crop in *j*th LDC can be estimated by using the following equation:

$$GAR_{i,j} = \sum_{i=j=1}^{m,n} \left(A_{i,j}Y_{i,j}\right)RPR_{i,j} \quad (1)$$

where $A_{i,j}$ and $Y_{i,j}$ respectively represent the area and yield of *i*th crop (*i* = 1, 2, 3, ... m crop) in the *j*th LDC (*j* = 1, 2, 3, ... n LDCs) and $RPR_{i,j}$ the residue to product ratio for *i*th crop in *j*th LDC.

The surplus agricultural residue available for bioelectricity can be evaluated by introducing certain restrictions on the GAR potential of the crop residues. It has been revealed that the competing uses of a particular crop residue and the harvesting practices have a remarkable influence on the availability of crop residues [55,56]. Moreover, a certain amount of crop residues is also required for retaining soil fertility [56–59].

Therefore, the surplus agricultural residue (*SAR*) availability of *i*th crop in *j*th LDC for energy applications can be estimated by using the following equation:

$$SAR_{i,j} = \xi_{ce}\left(1 - \xi_{fodder}\right)\left(1 - \xi_{oth}\right) \sum_{i=j=1}^{m,n} \left(A_{i,j}Y_{i,j}\right)RPR_{i,j} \quad (2)$$

where ξ_{ce} represents the collection efficiency of agricultural residues, ξ_{fodder} the fraction of agricultural residues used for fodder applications, and ξ_{oth} the fraction of agricultural residues used for other applications (i.e., paper industry, cardboard industry, construction materials, etc.).

Finally, the bioelectricity potential is estimated as a product of the surplus agricultural residue availability (Mt) for biomass power and specific biomass consumption for electricity (kg/kWh). It should be noted that not all agricultural residues are easily accessible, available, or economically viable for energy production [60]. Several factors are required to help determine the extent to which agricultural residues can be extracted in an environmentally benign manner from any specific location [61]. This includes, for example, crop cutting height, crop yield, land slope, tillage, edaphic factors (i.e., soil type and soil fertility), weather, and wind patterns [20,62,63]. Based in part on the recommendation by OECD/IEA [64], this study adopts a rather conservative 20% extraction rate for agricultural residues (leaving the remaining 80% for soil nutrient recycling, ecosystem function, animal fodder, and other competing utilization). The authors recommend that further to their results, geospatial analysis of crop production and robust field studies in situ would be required to inform policymakers regarding the realistic potential of agricultural residues that can be extracted in an environmentally benign manner in any of the studied countries. Additionally, the study helps provide invaluable information on specific crop type(s) to concentrate edaphoclimatic investigations on residues for future bioelectricity production. As reported in Ackom et al. [20], such information is essential to help address both food security and modern energy needs (via bioelectricity from only 20% agricultural residues) in developing countries.

The historical crop production data of major crops by the LDCs considered in this study has been obtained from FAO-Statistics [13]. Table S8a presents the production of major crops by country in 2017 obtained from FAO-Statistics [13]. The specific ratios of residue-to-grain production of different crops are taken from publicly available literature, as shown in the supplemental Table S8b. To assess the technical potential of agricultural residue availability we assume that the gross residue available from the crop production is available for bioelectricity. For the year 2017, the total crop production was 72.9 Mt in Bangladesh, 20.6 Mt in Ethiopia, 15.9 Mt in Nepal, 9.9 Mt in Zambia, 18.2 Mt in Malawi, and 10.4 Mt in Lao-PDR, as shown in the supplement (Table S8a). The gross residue availability was estimated at 111 Mt in Bangladesh, 34 Mt in Ethiopia, 21 Mt in Nepal, 15 Mt in Lao-PDR, 13 Mt in Zambia, and 17 Mt in Malawi for 2017.

We use a linear regression model to estimate the area and production of major crops in the near future. It is a starting point for projecting the agricultural residues and their utilization. For the years 2018 and 2030, the projected crop productivity is based on the data from 2002 to 2017 [13]. We assume that the increase in food production and productivity would follow the trend (2002–2017) for another 10–15 years, i.e., until 2030. In addition, there would no drastic increase or change in agricultural practices and/or shift in agricultural commodities. Figure 8 presents the historical and projected technical potential of agricultural residues by crop in the selected LDCs considered in this study. The annual technical potential of agricultural residue availability for bioelectricity and associated biomass power potential for six countries is presented in Table S9a of the supplementary information.

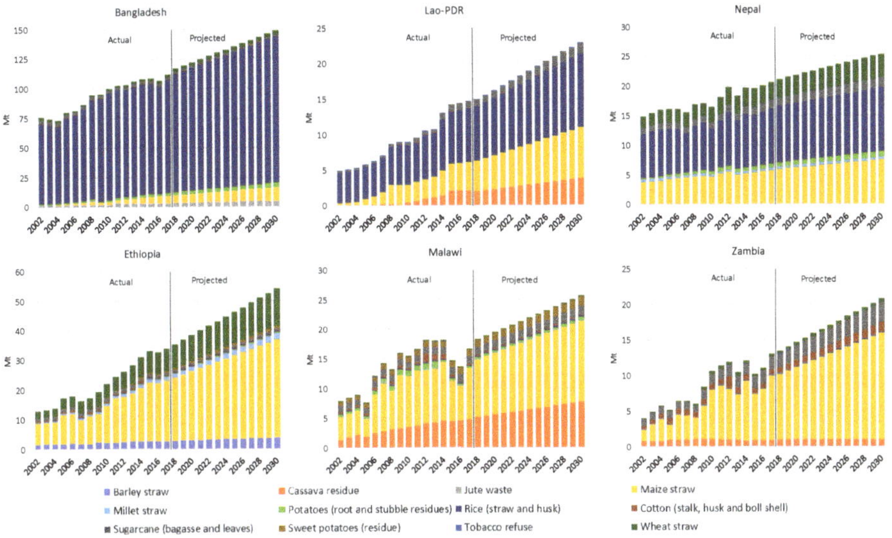

Figure 8. Historical and projected technical potential of agricultural residues by crop in the selected LDCs.

Figure 9 presents the technical potential of gross agricultural residue availability for biomass power generation in the selected LDCs considered in this study. The gross agricultural residue availability for energy applications is presented in the left panel whereas the technical potential of biomass power is shown in the right panel. The specific biomass consumption has been taken to be 1.21 kg/kWh with a capacity factor of 80% [65], whereas 1.6 kg/kWh specific bagasse consumption is assumed for bagasse-based co-generation units with a capacity factor of 53% [66]. For the base year 2017, the gross agricultural residue availability estimated at 111.4 Mt for Bangladesh is expected to increase 148.9 Mt by 2030. Using the assumptions on specific biomass/bagasse consumption, the technical potential of biomass power generation is estimated approximately at 13.2 and 17.6 GW, respectively, in 2017 and 2030. For Ethiopia, the gross agricultural residue availability estimated at 33.8 Mt in 2017 is expected to increase 54.1 Mt by 2030 (Figure 9). The associated technical potential of biomass power is estimated approximately at 4.0 and 6.4 GW, respectively, in 2017 and 2030. Similarly, in Lao-PDR, the gross agricultural residue availability estimated at 14.6 Mt in 2017 is expected to increase 22.9 Mt by 2030, whereas the technical potential of biomass power is estimated approximately at 1.7 and 2.7 GW, respectively, in 2017 and 2030. In Malawi, the gross agricultural residue availability estimated at 16.5 Mt in 2017 is expected to increase 25.5 Mt by 2030, whereas the technical potential of biomass power is estimated approximately at 2.0 and 3.0 GW, respectively, in 2017 and 2030 (Figure 9). In Nepal, the gross agricultural residue availability estimated at 20.6 Mt in 2017 is expected to increase 25.4 Mt by 2030, whereas the technical potential of biomass power is estimated approximately at 2.5 and

3.0 GW, respectively, in 2017 and 2030 (Figure 9). For Zambia, the gross agricultural residue availability estimated at 13.0 Mt in 2017 is expected to increase 20.8 Mt by 2030, whereas the technical potential of biomass power is estimated approximately at 1.6 and 2.5 GW, respectively, in 2017 and 2030 as shown in Figure 9 (see Table S9a).

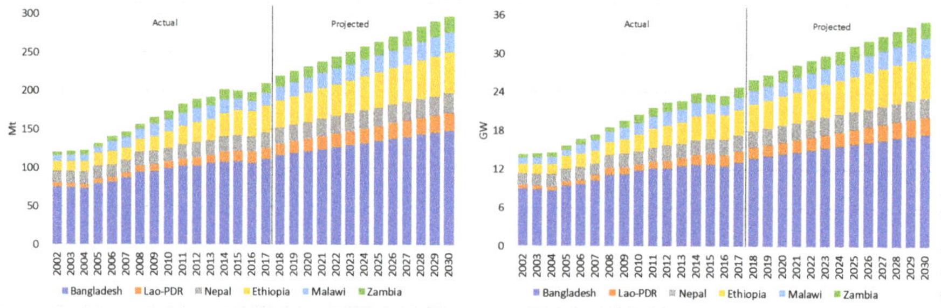

(a) **(b)**

Figure 9. Technical potential of agricultural residues for biomass power generation in the selected LDCs: (**a**) Agricultural residue availability for energy applications, (**b**) Biomass power generation potential.

The use of crop residues varies from region to region and depends on their calorific values, lignin content, density, palatability by livestock, and nutritive value. The residues of most of the cereals and pulses have fodder value. However, the woody nature of the residues of some crops restricts their utilization to fuel uses only. The dominant end uses of crop residues in the LDCs are as fodder for cattle, fuel for cooking, and thatch material for housing [17,67–69]. All the non-fodder, non-fertilizer agricultural residues with low moisture content can, in principle, be considered as feedstocks for energy applications. Approximately, 10% of the total amount of agricultural residues is lost in the collection, transportation, and storage, whereas another 15% of the total amount of agricultural residues is used in other competing applications [54]. A recent study by Purohit and Fischer [68] estimated that approximately 53% of the gross residue availability is used for fodder applications in India. Therefore, as mentioned above, to assess the economic potential of agricultural residues in the LDCs in this study we have assumed a conservative estimate of 20% of the gross residue availability (technical potential) used for energy applications. Figure 10 presents the historical and projected the economic potential of agricultural residues by crop in selected LDCs considered in this study. The annual economic potential based on surplus agricultural residue availability for bioelectricity and associated biomass power potential for six LDCs considered in this study is presented in Table S9b of the supplementary information.

Figure 11 presents the economic potential of surplus agricultural residue availability for biomass power generation in the selected LDCs considered in this study. The surplus agricultural residue availability for energy applications is presented in the left panel whereas the economic potential of biomass power is shown in the right panel. For the base year 2017, the surplus agricultural residue availability was estimated at 22.3 Mt for Bangladesh is expected to increase 29.8 Mt by 2030. The economic potential of biomass power generation is estimated approximately at 2.6 and 3.5 GW, respectively, in 2017 and 2030. For Ethiopia, the surplus agricultural residue availability was estimated at 6.8 Mt in 2017 is expected to increase by 10.8 Mt by 2030. The associated technical potential of biomass power is estimated approximately at 0.8 and 1.3 GW, respectively, in 2017 and 2030. Similarly, in Lao-PDR the surplus agricultural residue availability was estimated at 2.9 Mt in 2017 is expected to increase approximately 4.6 Mt by 2030, whereas the technical potential of biomass power is estimated approximately at 0.3 and 0.5 GW, respectively, in 2017 and 2030. In Malawi, the surplus agricultural residue availability was estimated at 3.3 Mt in 2017 is expected to increase 5.1 Mt by 2030, whereas the technical potential of biomass power is estimated approximately at 0.4 and 0.6 GW, respectively, in 2017 and 2030. In Nepal, the surplus agricultural residue availability was estimated at 4.1 Mt in 2017 is expected to increase 5.1 Mt by 2030, whereas the technical potential of biomass power is estimated approximately at 0.5 and 0.6 GW, respectively, in 2017 and 2030. For Zambia, the surplus agricultural residue availability was estimated at 2.6 Mt in 2017 is expected to increase about 4.2 Mt by 2030, whereas the technical potential of biomass power is estimated approximately at 0.3 and 0.5 GW, respectively, in 2017 and 2030 as shown in Figure 11 (see Table S9b).

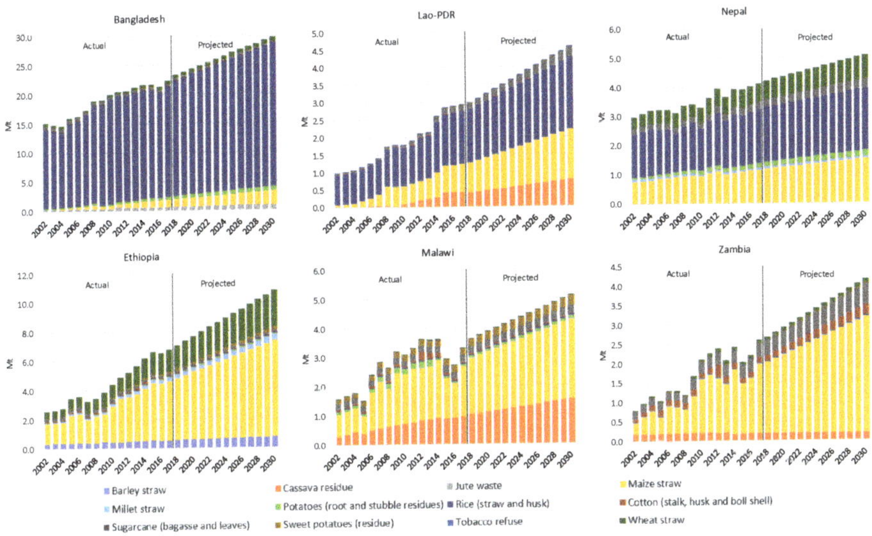

Figure 10. Historical and projected economic potential of agricultural residues by crop in the selected LDCs.

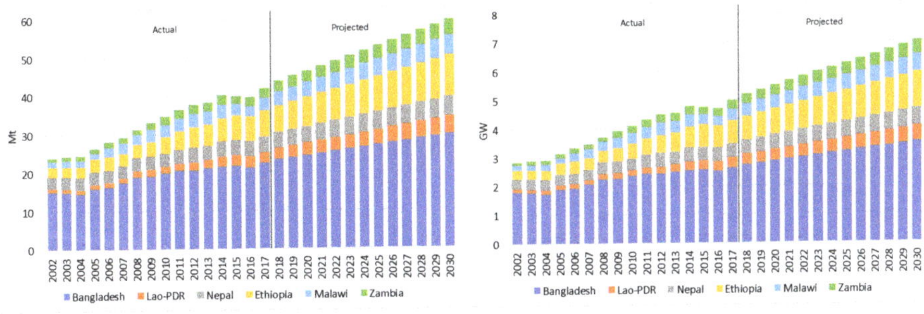

(a) (b)

Figure 11. Economic potential of agricultural residues for biomass power generation in the selected LDCs: (**a**) Agricultural residue availability for energy applications, (**b**) Biomass power generation potential.

The prevailing practices of using agricultural residues for alternative applications (fuel, fodder, construction materials, cardboard industry, etc.) varied across the countries analyzed in this study. Therefore, we have carried out a sensitivity analysis to show the impact of using different shares of surplus agricultural residues on biomass availability for energy applications (Figure 12). As expected, surplus biomass feedstock availability and associated bioelectricity potential are highly sensitive to the collection, transportation, and storage losses, and other prevailing practices of agricultural residue used for several end-use applications across the countries analyzed in this study.

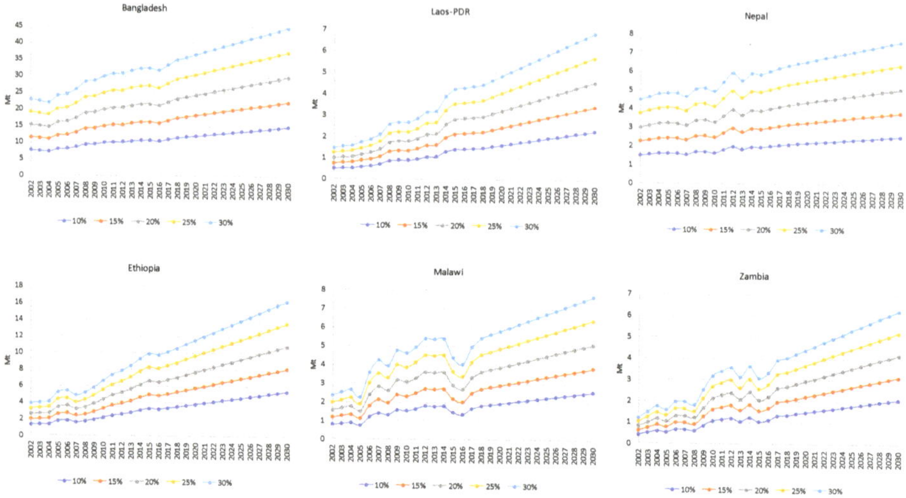

Figure 12. Impact of using different shares of surplus agricultural residues on biomass availability for energy applications.

5. Electricity Supply and the Role of Bioelectricity for Meeting the SDG7

5.1. *Technical and Economic Potential of Bioelectricity in the Selected LDCs*

Using the input parameters and assumptions mentioned above (Section 4), Figure 13 presents the technical and economic potential (in TWh) of electricity generation based on gross (technical) and net (economic) agricultural residue availability in selected LDCs. Table S9c presents the technical and economic potential of agricultural residues for bioelectricity in the selected LDCs.

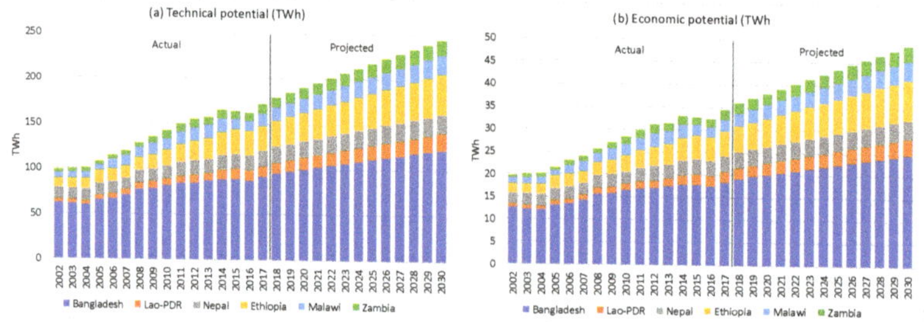

Figure 13. Technical and economic potential (in TWh) of electricity generation by surplus biomass obtained from the agriculture sector in the selected LDCs analyzed in this study.

Table 4 presents the economic potential of biomass power (MW) and its contribution (% share) in peak power demand of the selected LDCs. In Malawi, bioelectricity contributed around 92% of the total installed capacity, followed by Nepal (30%), Bangladesh (24%), Lao-PDR (22%), Ethiopia (21%), and Zambia (12%) in 2017 (see Table 4). Table 5 provides the economic potential of bioelectricity (GWh) and its contribution to the total electricity demand of the selected LDCs. It is observed that in the base year 2017, bioelectricity obtained from surplus agricultural residues provided 110% of the total electricity demand in Malawi, followed by Nepal (45%), Bangladesh (29%), Lao-PDR (29%), Ethiopia (27%), and Zambia (13%). As expected, the share of bioelectricity will decrease in the near future primarily due to the rapidly increasing demand for electricity in the selected LDCs (see Table 5).

Table 4. Total peak demand, biomass power potential, and its contribution (% share) in selected years.

Year	Total Peak Power (MW)						Biomass Power Potential (MW)						% Share of Biomass Power					
	Bangladesh	Lao-PDR	Nepal	Ethiopia	Malawi	Zambia	Bangladesh	Laos	Nepal	Ethiopia	Malawi	Zambia	Bangladesh	Laos	Nepal	Ethiopia	Malawi	Zambia
2015	9036	1056	1292	2657	352	2504	2564	336	464	781	348	245	28.4	31.9	35.9	29.4	98.9	9.8
2016	9479	1349	1468	3156	388	2574	2513	341	478	770	323	265	26.5	25.3	32.5	24.4	83.1	10.3
2017	10,958	1608	1644	3748	427	2647	2634	347	491	799	394	311	24.0	21.6	29.9	21.3	92.3	11.8
2020	13,746	2723	2638	6279	571	2893	2879	385	519	907	461	351	20.9	14.1	19.7	14.4	80.8	12.1
2025	20,056	4395	4519	9989	925	3401	3196	465	562	1092	534	425	15.9	10.6	12.4	10.9	57.7	12.5
2030	29,264	5892	7542	14,372	1500	4066	3514	545	604	1277	607	499	12.0	9.3	8.0	8.9	40.5	12.3

Source: authors' own estimates.

Table 5. Electricity demand, bioelectricity potential, and its contribution (% share) in the total electricity demand, selected years.

Year	Electricity Demand (GWh)						Economic Potential of Bioelectricity (GWh)						% share of Bioelectricity					
	Bangladesh	Lao-PDR	Nepal	Ethiopia	Malawi	Zambia	Bangladesh	Laos	Nepal	Ethiopia	Malawi	Zambia	Bangladesh	Lao-PDR	Nepal	Ethiopia	Malawi	Zambia
2015	52,193	5212	6335	14,637	1997	15,355	17,852	2306	3177	5438	2367	1614	34.2	44.2	50.1	37.2	118.5	10.5
2016	57,276	6789	6912	17,415	2208	15,784	17,501	2340	3267	5357	2187	1748	30.6	34.5	47.3	30.8	99.0	11.1
2017	62,678	8188	7490	20,720	2432	16,231	18,358	2378	3358	5567	2687	2069	29.3	29.0	44.8	26.9	110.5	12.7
2020	79,533	14,378	12,018	34,906	3249	17,740	20,092	2636	3553	6319	3154	2333	25.3	18.3	29.6	18.1	97.1	13.1
2025	118,288	24,057	20,585	53,132	5268	20,855	22,339	3182	3843	7618	3659	2829	18.9	13.2	18.7	14.3	69.5	13.6
2030	175,926	32,923	34,355	73,709	8541	24,933	24,588	3728	4132	8917	4165	3326	14.0	11.3	12.0	12.1	48.8	13.3

Source: authors' own estimates.

5.2. Achieving Universal Access to Electricity (SDG7.1) in the Rural Households in the LDCs

As mentioned above, access to electricity is critically important for economic development in the LDCs. SDG7.1 (Indicator 7.1.1) aims to achieve affordable, reliable, and modern electricity access at the household level by 2030 (see Table S3). Electricity consumption (kWh/capita) and HDI score (development indicators) are proportionally related, i.e., higher electricity consumption leads to the higher development and human welfare indicators [70,71]. Rural households in the LDCs lack electricity access and it is important to provide basic electricity services. This should be affordable since the households cannot spend a lot of their monthly expenditure on purchasing electricity [72]. The "Energy for All" initiative aims at providing electricity access at the initial level of 250 kWh per rural household [72]. Brecha [70,71] argues that access to sufficient amounts of electricity is a prerequisite for social well-being and proposes an electricity access threshold for meeting non-energy SDGs at 400 kWh per capita, i.e., less than this average per capita threshold may lead to the poor HDI score.

The World Bank developed a framework/matrix for measuring access to household supply [73] and the UNDP recently reflected on how the electricity consumption is related to the SDG1, SDG3, SDG4, SDG5 using the five tiers of access to electricity [74]. The first scenario or level of electricity consumption would be "Tier 1," i.e., lighting and phone charging. The indicative consumption of electricity per household in the five tiers is presented in Table 6. If we consider lighting, phone charging, television, and medium power appliances, the total annual electricity consumption per household would be more than 365 kWh [73].

Table 6. Multi-tier matrix for household electricity consumption.

Particulars	Tier 1	Tier 2	Tier 3	Tier 4	Tier 5
Tier Criteria	Task lighting + phone charging	**Tier 1** + Television + Fan (if needed)	**Tier 2** + any medium-power appliances	**Tier 3** + any high-power appliances	**Tier 4** + any very high-power appliances
Annual consumption levels (kWh)	≥4.5	≥73	≥365	≥1250	≥3000

Source: World Bank [73].

The average household size varies from 4–5 persons in the LDCs [75]. Table 7 provides the number of rural households in the LDCs and bioelectricity available for domestic uses. The study finds that bioelectricity can meet the Tier 3 electricity demand in the rural households (see Tables 6 and 7), except in Ethiopia. Lao-PDR has the highest bioelectricity supply, i.e., 522 kWh/capita, which crosses the 400 kWh threshold for meeting the SDGs, followed by Zambia (211 kWh/capita), Malawi (183 kWh/capita), Bangladesh (179 kWh/capita), Nepal (151 kWh/capita), and Ethiopia (66 kWh/capita).

Table 7. Bioelectricity availability for domestic (household) uses in rural areas in 2017.

Country	Rural Population		Size of Household	Number of Rural Households (Thousand)	Bioelectricity Potential (GWh)	Bioelectricity (kWh) *	
	(Million)	% of Total				Per Household	Per Capita
Bangladesh	102.3	64.1	4.5	22,744	18,358	807	179
Lao-PDR	4.6	65.6	5.3	860	2378	2764	522
Nepal	22.3	80.7	4.4	5068	3358	663	151
Ethiopia	84.8	79.7	4.6	18,435	5567	302	66
Malawi	14.7	83.3	4.5	3271	2687	821	183
Zambia	9.8	57	5.1	1923	2069	1076	211

* Available for domestic and other productive uses. Source: Authors' estimates.

5.3. Strengthening Interlinkages: Modern Bioenergy, SDG7, and Other SDGs

As mentioned above, low-income countries in SSA and Asia are predominantly using traditional biomass in inefficient devices for cooking and heating purposes. Modern bioenergy has a potentially positive impact on the fulfillment of the SDGs, ranging from SDG7 (increase share of RE) to SDG13 (mitigation of climate change), SDG1 (no poverty) and SDG8 (on poverty and economic growth, e.g., by enhancing the competitiveness of agro-industries, thus contributing to job creation and social welfare). It is important to discuss the interlinkages between modern bioenergy production and targets of SDGs. Recently, Fritsche et al. [76] presented the interlinkages between the SDGs and the Sustainability Indicators for Bioenergy (GSI). Souza et al. [77] highlighted the role of bioenergy in a climate-changing world. Synergies on food security and bioenergy, improvement in resource efficiency, and integrated assessment of bioenergy systems are needed for meeting the SDGs. IRENA (International Renewable Energy Agency 2017) has highlighted the constructive role of bioenergy in achieving the agreed UN SDGs and implementing the Paris Agreement on Climate Change [78]. This includes climate goals (SDG13), food security (SDG2), better land use (SDG15), and clean and sustainable energy (SDG7). A holistic approach is required to evaluate the performance of bioenergy supply chains for meeting SDGs, including energy and food security, and socio-economic development.

Rural electrification based on biomass power would greatly contribute to the welfare of rural people. The implications are obviously lighting, (also TV and radio), education, health, productivity increase, and environment [79]. Kerosene lamps are widely used for lighting the households in low-income countries. Around 53% of the population still use kerosene or oil lamps for lighting purposes in SSA (WHO, 2016) as there is no access to electricity. Small particulate matters (PM) emissions would be emitted while burning the lamps and they are associated with severe indoor health risks [80,81]. Lam et al. [82] observed that reductions to fine particulate matter ($PM_{2.5}$) resulting from reducing kerosene use for lighting in India could avert between 50 and 300 thousand adult DALYs (Disability Adjusted Life Years Lost) in 2030. Bioelectricity can help replace kerosene lamps, thereby improving indoor air quality and reducing associated health impacts and saving fossil fuel (kerosene).

When food systems are integrated with agricultural residues for energy production, there would be an increase in agricultural productivity and income of small-scale food producers (SDG2.3). As mentioned above, the open burning of agricultural residues would cause air pollution. Abandoned biomass residues are collected and utilized in agro-industries, such as sugarcane industries or rice mills, can help improve the competitiveness of agro-industries and the economic contribution of the agriculture sector would increase. Thus, biomass power would contribute to improving the productivity and incomes of small-scale food producers (SDG2.4).

If we use bioelectricity for productive use in SMEs (small and medium enterprises) or water pumping or to electricity in schools and hospitals for displacing diesel/back-up generator, there would be direct GHG savings in a range of 1.7 Mt (Lao-PDR) to 13.4 Mt (Bangladesh). Table 8 shows the potential diesel substitution and GHG emissions mitigation potential in the selected LDCs. Modern bioenergy contributes to reducing GHG emissions and brings added benefits [77]. Electricity supply might also contribute in increasing the crop yields (agricultural productivity) by irrigating farmlands using water-pumping schemes.

Table 8. Fuel savings and avoided global greenhouse gas (GHG) emissions from the substitution of diesel power generation by bioelectricity.

Particulars	Bangladesh	Lao-PDR	Nepal	Ethiopia	Malawi	Zambia
Total bioelectricity production (TWh)	18.4	2.4	3.4	5.6	2.7	2.1
Total diesel savings (million liters)	5038	6536	922	1528	737	568
GHG emissions avoided (million tonnes)	13.4	1.7	2.5	4.1	2.0	1.5

Source: Emission factor is taken from IPCC, 2006 [83].

5.4. Bioelectricity: Complementarity with Hydropower and Synergy with Agro-Industries

Nepal has only harnessed around 1.3% (i.e., 1051 MW hydropower from the total potential 83,000 MW) hydropower potential mainly due to the high cost of constructing large power plants, the scattered population in rural areas, and associated grid extension costs in remote and mountainous topography [84]. Therefore, distributed or isolated electricity systems such as small micro-hydro, solar, and bio-power (electricity from biomass) might be one of the reliable alternative options for rural electrification. The Government of Nepal (GoN) has enacted policy instruments (subsidies and tax benefits) to promote the RE (especially micro-hydro and solar photovoltaic systems) which has provided electricity to more than 12% of the country's population [42]. Most of the hydropower plants in Nepal are of run-of-rivers types [28,43]. This means that the water flow varies significantly between the wet and dry seasons. Therefore, electricity generation fluctuates and is highly seasonal. In the past, Nepal has faced an extended load-shedding [19], resulting in negative implications for economic development in the country. It is important to complement hydroelectricity by other sources of electricity. This not only helps to reduce the cost of electricity infrastructure but also to reduce the electricity imports and promote the rural development. Nepal has also planned to generate 220 MW of electricity from bioenergy by 2030 in its first NDC (Nationally Determined Contribution) report [85].

In Zambia, electricity generation has not increased in the last three years [41], mainly due to less water flows in rivers, leading to negative consequences on the nation's economic growth. The country also imports electricity from the neighboring countries (such as Mozambique and South Africa) depending upon the availability of water in rivers. For example, the country has imported 1391 GWh electricity in 2016 [41]. As mentioned earlier, Zambia's electricity supply is dominated by hydropower. Recently, Zambia has experienced an extended load-shedding (power-cuts) in the country, reducing its electricity consumption by 30% in 2015 [47]. This is mainly due to low water levels in reservoirs and declining water flows in rivers [46]. Thus, it is important to diversify electricity generation capacity for sustaining the current level of economic growth in order to meet SDGs.

In Malawi, electricity generation output reduced to 150 MW in December 2017, due to less water flow in the river, with more than two times reduction compared to its installed capacity [51]. This is to mention that 98% of this electricity comes from hydropower plants installed along the Shire River [45]. The installed capacity for electricity generation is lower than the demand. Thus, biomass-based electricity could enhance the security of electricity supply in Malawi. Furthermore, the household or domestic sector consumes a large amount of electricity and peak demand occurs in the morning and evening. For example, in Zambia the highest power demand (peak demand) happens between 7:00–10:00 a.m. and 4:00–9:00 p.m. [52]. Therefore, bioelectricity from agriculture residues can contribute to both baseload and peak-load power in the LDCs.

How much biomass power can contribute to meeting the existing domestic consumption and saving the import of electricity from the neighboring countries? Bioelectricity could cover 57% of the domestic energy consumption in Bangladesh, 90% in Ethiopia, and 43% in Zambia. Interestingly, bioelectricity is more than the domestic demand in Nepal, Lao-PDR, and Malawi (Table 9). On the other hand, imported electricity can be fully substituted by bioelectricity in Bangladesh, Lao-PDR, and Nepal.

Table 9. Domestic energy consumption, imports, and bioelectricity potential (TWh)*.

Country	Domestic Energy Consumption (TWh)	Electricity Imports (TWh)	Bioelectricity Potential (TWh)	Reference Year
Bangladesh	29.1	4.8	18.4	2017
Lao-PDR	1.6	0.8	2.3	2015
Nepal	1.8	2.2	3.4	2017
Ethiopia	5.8	-	5.4	2016
Malawi	0.9	-	2.2	2016
Zambia	4.4	2.2	1.8	2016

* Note: Domestic energy consumption and electricity imports are obtained from [48] for Bangladesh; [9] for Lao-PDR; [43] for Nepal, and [41] for Ethiopia, Malawi, and Zambia.

Electricity production from agricultural residues would help in improving the economic competitiveness of agro-industries such as sugarcane mills and rice mills. Surplus electricity from the mills can be sold to the grid. This needs high attention from policymakers since agricultural productivity has been stagnating in many other developing countries and the sector's contribution to GDP is decreasing across the LDCs analyzed in this study.

5.5. *Realizing the Bioelectricity Potential—A Way Forward*

There are several technologies for energy generation from biomass [19,65,86–89], mainly combustion (heat and power plants) and gasification (gasifier coupled with gas or diesel engine). Biomass combustion is a widely used technology for power generation from agricultural residues and biomass co-products [19]. Recently, electricity from biomass gasification has fast emerged for rural electrification using agricultural residues in developing countries [87]. Rice- and straw-based biomass power plants can be integrated with other electricity sources and the utility grid [90]. A few novel technologies on efficient conversion of biomass are being investigated [91], which may help in diversifying the energy systems using different pathways.

Biomass gasification generates producer gas (a mixture of combustible gas) which can be later combusted in a diesel engine (in dual mode) or cent percent (100%) gas engines for producing electricity. Thus, it is suitable for distributed or off-grid power for rural areas. We need to scrutinize several factors such as the price of biomass, investments, logistics costs, electricity price, environmental constraints, and regulatory uncertainties for making the biomass gasifiers cost-competitive [89,90,92,93].

Despite government policies on modernizing bioenergy in the LDCs, there is a lack of investments in biomass power projects. Therefore, governments should make efforts towards creating a favorable environment for electricity generation from agricultural residues, ranging from policy frameworks, financing, and business models. The utilization of agricultural residues for decentralized power generation could also provide financial incentives to the farmers if the pricing system for agricultural biomass is well designed. The involvement of local agencies and stakeholders (farmers, agro-industries, and investors) is also important in a successful implementation of efficient biomass power technologies. A synergetic approach with other sectors of the economy such as agriculture and industry would help in mobilizing the stakeholders and resources, especially investments.

6. Conclusions

Least Developed Countries (LDCs) still do not have full access to modern energy services (electricity and clean cooking fuel), especially in rural areas. The utilization of conservative and environmentally benign amounts of only 20% of surplus agricultural residues for bioelectricity could help complement the grid regarding baseload electricity supply as well as provision of the service to rural households via decentralized systems in developing countries. Bioelectricity can provide power to small and medium enterprises (e.g., agro-industries) and help in modernizing agricultural systems such as irrigation and tillage operation, thereby improving the agricultural yield.

Modern bioenergy would help in creating synergies with other sectors of the economy such as agriculture, industry, and rural development. The productive uses of bioelectricity in agro-industries can substitute backup fossil-based generators. Based on findings, surplus agricultural residues could provide the total current electricity demand in Malawi alone, followed by Nepal (45%), Bangladesh (29%), Lao-PDR (29%), Ethiopia (27%), and Zambia (13%). This study also explores the complementarity and synergies of bioelectricity, SDG7, and their interlinkages with other SDGs. Bioelectricity could potentially improve ambient air quality and improve public health (SDG3) and reduce GHG emissions (SDG13) as well. It is recommended that the transition to bioelectricity needs to be planned together with investment plans, infrastructure, linkages with other sustainable development goals (SDGs), and engagement with relevant stakeholders (investors and policymakers). The established positive link with the other SDGs would help expedite the process of transitions towards the effective use of bioelectricity in the LDCs.

Technologies for biomass to bioelectricity conversion are mature. As part of the global climate agenda for reducing GHG emissions, promotion of sustainable RE technologies, and universal access to electricity, international cooperation might be sought to attract financing for a successful implementation of bioelectricity projects. This will help promote clean electricity using local indigenous resources. Hence, the management of biomass logistics and financing options should be further explored while harnessing the bioelectricity potential. A systems approach in policy design and close cooperation with public and private sectors is needed together with the engagement of farmers and communities in energy planning at the local level. Therefore, bioelectricity holds the potential to provide a catalytic role in mobilizing efforts to create a new impulse in national economies in the studied LDCs and other developing countries.

Supplementary Materials: The following are available online at http://www.mdpi.com/2071-1050/11/24/7091/s1, Figure S1. LDCs analyzed in the present study; Figure S2. GDP growth rate in the selected LDCs; Figure S3. Population and urbanization pattern in the selected LDCs; Figure S4. Electricity access from 1990 to 2016 in the world, South Asia, SSA and selected LDCs; Table S1. Total Primary Energy Supply (TPES) in the LDCs by fuel source (in ktoe); Table S2. Land covered by agricultural and forest land (% of the country's land area); Table S3. Targets and indicators for measuring SDG7; Table S4. Access to electricity—Indicator 7.1.1 in the rural and urban population in LDCs, 2000–2016, selected years; Table S5. Status of SDG7—Indicators: clean cooking (7.1.2), share of renewable energy (7.2.1) and primary energy intensity (7.3.1) in the LDCs (in 2016); Table S6. Primary energy Primary energy intensity (MJ per 2011 US$ PPP) in the selected LDCs; Table S7: Projection of peak load (load capacity) and electricity generation in the LDCs; Table S8a: Production of major crops by country in 2017; Table S8b: Residue to grain ratio for major crops in the LDCs; Table S9a: Technical potential of agricultural residues for bioelectricity; Table S9b: Economic potential of agricultural residues for bioelectricity; Table S9b: Economic potential of agricultural residues for bioelectricity; Table S9c: Technical and economic potential of agricultural residues for bioelectricity.

Author Contributions: Conceptualization, D.K., P.P., and E.K.A.; methodology, D.K., and P.P.; writing—original draft preparation, D.K., and P.P.; writing—review and editing, D.K., P.P., and E.K.A.

Funding: This research received no external funding.

Conflicts of Interest: The authors declare no conflict of interest.

Disclaimer: The opinions and recommendations expressed in this entry are those of the authors and do not necessarily reflect the views of KTH Royal Institute of Technology, International Institute for Applied Systems Analysis (IIASA) and/or UNEP DTU Partnership. The designations employed and the presentation of the material in this publication do not imply the expression of any opinion whatsoever on the part of KTH Royal Institute of Technology, International Institute for Applied Systems Analysis (IIASA) and/or UNEP DTU Partnership concerning the legal status of any country, territory, city, or area or of its authorities. References to different sources have been made in this document.

References

1. UN (United Nations). About the Sustainable Development Goals—United Nations Sustainable Development. 2015. Available online: https://www.un.org/sustainabledevelopment/sustainable-development-goals/ (accessed on 24 September 2019).
2. UNDP Goal 7: Affordable and Clean Energy. Available online: https://www.undp.org/content/undp/en/home/sustainable-development-goals/goal-7-affordable-and-clean-energy.html (accessed on 24 September 2019).
3. UNCTAD. Chapter 2: Energy and inclusive economic structural transformation. In *The Least Developed Countries Report 2017 (Transformational Energy Access)*; United Nations Conference on Trade and Development (UNCTAD): Geneva, Switzerland, 2017.
4. UN (United Nations). LDC Identification Criteria & Indicators|Department of Economic and Social Affairs. 2019. Available online: https://www.un.org/development/desa/dpad/least-developed-country-category/ldc-criteria.html (accessed on 24 September 2019).
5. Goldemberg, J.; Martinez-Gomez, J.; Sagar, A.; Smith, K.R. Household air pollution, health, and climate change: Cleaning the air. *Environ. Res. Lett.* **2018**, *13*, 030201. [CrossRef]
6. Yip, F.; Christensen, B.; Sircar, K.; Naeher, L.; Bruce, N.; Pennise, D.; Lozier, M.; Pilishvili, T.; Loo Farrar, J.; Stanistreet, D.; et al. Assessment of traditional and improved stove use on household air pollution and personal exposures in rural western Kenya. *Environ. Int.* **2017**, *99*, 185–191. [CrossRef] [PubMed]
7. Das, I.; Jagger, P.; Yeatts, K. Biomass cooking fuels and health outcomes for women in Malawi. *HHS Public Access* **2017**, *25*, 1032–1057. [CrossRef] [PubMed]
8. Taulo, J.L.; Gondwe, K.J.; Sebitosi, A.B. Energy supply in Malawi: Options and issues. *J. Energy South. Africa* **2015**, *26*, 19–32. [CrossRef]
9. Ministry of Energy and Mines. *Lao PDR Energy Statistics 2018*; Ministry of Energy and Mines: Vientiane, Lao People's Democratic Republic (Lao PDR), 2018.
10. IEA. "IEA Statistics" IEA—International Energy Agency. Available online: https://www.iea.org/statistics (accessed on 24 September 2019).
11. World Bank. "World Bank Open Data—Statistics". Available online: https://data.worldbank.org (accessed on 24 September 2019).
12. UNCTAD. *Statistical Tables on the Least Developed Countries—2017*; United Nations Conference on Trade and Development (UNCTAD): Geneva, Switzerland, 2017.
13. FAO-STAT. Crops: "FAO - Food and Agriculture Organization of the United Nations". Available online: http://www.fao.org/faostat/en/#data/QC (accessed on 24 September 2019).
14. Huda, A.S.N.; Mekhilef, S.; Ahsan, A. Biomass energy in Bangladesh: Current status and prospects. *Renew. Sustain. Energy Rev.* **2014**, *30*, 504–517. [CrossRef]
15. Halder, P.K.; Paul, N.; Beg, M.R.A. Assessment of biomass energy resources and related technologies practice in Bangladesh. *Renew. Sustain. Energy Rev.* **2014**, *39*, 444–460. [CrossRef]
16. Toth, G.G.; Nair, P.K.R.; Jacobson, M.; Widyaningsih, Y.; Duffy, C.P. Malawi's energy needs and agroforestry: Impact of woodlots on fuelwood sales. *Energy Sustain. Dev.* **2019**, *50*, 101–108. [CrossRef]
17. Shane, A.; Gheewala, S.H.; Fungtammasan, B.; Silalertruksa, T.; Bonnet, S.; Phiri, S. Bioenergy resource assessment for Zambia. *Renew. Sustain. Energy Rev.* **2016**, *53*, 93–104. [CrossRef]
18. Gurung, A.; Oh, S.E. Conversion of traditional biomass into modern bioenergy systems: A review in context to improve the energy situation in Nepal. *Renew. Energy* **2013**, *50*, 206–213. [CrossRef]
19. Khatiwada, D.; Seabra, J.; Silveira, S.; Walter, A. Power generation from sugarcane biomass—A complementary option to hydroelectricity in Nepal and Brazil. *Energy* **2012**, *48*, 241–254. [CrossRef]
20. Ackom, E.K.; Alemagi, D.; Ackom, N.B.; Minang, P.A.; Tchoundjeu, Z. Modern bioenergy from agricultural and forestry residues in Cameroon: Potential, challenges and the way forward. *Energy Policy* **2013**, *63*, 101–113. [CrossRef]
21. Sasaki, N.; Knorr, W.; Foster, D.R.; Etoh, H.; Ninomiya, H.; Chay, S.; Kim, S.; Sun, S. Woody biomass and bioenergy potentials in Southeast Asia between 1990 and 2020. *Appl. Energy* **2009**, *86*, S140–S150. [CrossRef]
22. Kowsari, R. Three dimensional energy profile:: A conceptual framework for assessing household energy use. *Energy Policy* **2011**, *39*, 7505–7517. [CrossRef]
23. Kander, A.; Stern, D.I. Economic growth and the transition from traditional to modern energy in Sweden. *Energy Econ.* **2014**, *46*, 56–65. [CrossRef]

24. UNCTAD. *Statistical Tables on the least Developed Countries (LDCs)—2018*; UNCTAD: Geneva, Switzerland, 2018.
25. UNDP. (United Nations Development Programme)|Human Development Reports. Available online: http://hdr.undp.org/en/composite/HDI (accessed on 24 September 2019).
26. BP Statistical Review of World Energy. Available online: https://www.bp.com/en/global/corporate/energy-economics/statistical-review-of-world-energy/downloads.html (accessed on 16 October 2019).
27. Global Carbon Atlas CO2 Emissions|Global Carbon Atlas. Available online: http://www.globalcarbonatlas.org/en/CO2-emissions (accessed on 10 August 2019).
28. GoN Water and Energy Commission Secretariat. *Electricity Demand Forecast Report (2015-2040)*; Government of Nepal (GoN): Kathmandu, Nepal, 2017. Available online: https://www.wecs.gov.np/uploaded/Electricity-Demand-Forecast-Report-2014-2040.pdf (accessed on 14 July 2019).
29. UNDP. *UNDP (United Nations Development Programme) Energy Profile of Malawi*; UNDP: New York, NY, USA, 2016.
30. IEA. Energy Access Outlook. 2017. Available online: https://www.iea.org/access2017/ (accessed on 24 September 2019).
31. Chen, J.; Li, C.; Ristovski, Z.; Milic, A.; Gu, Y.; Islam, M.S.; Wang, S.; Hao, J.; Zhang, H.; He, C.; et al. Science of the total environment a review of biomass burning: Emissions and impacts on air quality, health and climate in China. *Sci. Total Environ.* **2017**, *579*, 1000–1034. [CrossRef] [PubMed]
32. Bhuvaneshwari, S.; Hettiarachchi, H.; Meegoda, J. Crop residue burning in India: Policy challenges and potential solutions. *Int. J. Environ. Res. Public Health* **2019**, *16*, 832. [CrossRef] [PubMed]
33. UNEP. *Air Pollution in Asia and the Pacific: Science-Based Solutions*; UNEP: Paris, France, 2019.
34. Khatiwada, D. *Assessing the Sustainability of Bioethanol Production in Different Development Contexts: A Systems Approach*; KTH Royal Institute of Technology: Stockholm, Sweden, 2013; ISBN 9789175018232.
35. FAOSTAT (land use indicator). Food and Agriculture Organization of the United Nations. 2019. Available online: http://www.fao.org/faostat/en/#data/EL (accessed on 24 September 2019).
36. Van der Kroon, B.; Brouwer, R.; van Beukering, P.J.H. The energy ladder: Theoretical myth or empirical truth? Results from a meta-analysis. *Renew. Sustain. Energy Rev.* **2013**, *20*, 504–513. [CrossRef]
37. Sovacool, B.K. Conceptualizing urban household energy use: Climbing the "Energy Services Ladder". *Energy Policy* **2011**, *39*, 1659–1668. [CrossRef]
38. World Bank. "Tracking SDG7—The Energy Progress Report". Available online: https://trackingsdg7.esmap.org/countries (accessed on 15 October 2019).
39. Quinn, A.K.; Puzzolo, E.; Dickinson, K.; Sturke, R.; Jack, D.W.; Mehta, S.; Shankar, A.; Sherr, K.; Rosenthal, J.P. An analysis of efforts to scale up clean household energy for cooking around the world. *Energy Sustain. Dev.* **2018**, *46*, 1–10. [CrossRef] [PubMed]
40. BPDB. *Annual Report 2016-17, Bangladesh Power Development Board*; Government of Bangladesh: Dhaka, Bangladesh, 2017; Volume 80.
41. UN-Stat United Nations, Statistics Division. Energy Statistics. Electricity Profiles. Available online: https://unstats.un.org/unsd/energystats/pubs/eprofiles/ (accessed on 25 September 2019).
42. AEPC. *Progress at a Glance: A Year in Review*; AEPC: Patan, Kathmandu, Nepal, 2018. Available online: https://www.aepc.gov.np/uploads/docs/progress-at-glance-a-year-in-review-fy-207475-201718-1541497047.pdf (accessed on 15 October 2019).
43. NEA. *Nepal Electricity Authority. 2018. A Year in Review: Fiscal Year 2017/2018*; NEA: Kathmandu, Nepal, 2018.
44. The Himalayan Times Diesel Generator Emissions Making Valley Air More Toxic. Available online: https://thehimalayantimes.com/kathmandu/diesel-generator-emissions-making-valley-air-more-toxic/ (accessed on 14 October 2019).
45. Zalengera, C.; Blanchard, R.E.; Eames, P.C.; Juma, A.M.; Chitawo, M.L.; Gondwe, K.T. Overview of the Malawi energy situation and A PESTLE analysis for sustainable development of renewable energy. *Renew. Sustain. Energy Rev.* **2014**, *38*, 335–347. [CrossRef]
46. Sladoje, M. Lights Out: Zambia's Electricity Crisis—IGC. Available online: https://www.theigc.org/blog/lights-out-zambias-electricity-crisis/ (accessed on 25 September 2019).
47. Batidzirai, B.; Moyo, A.; Kapembwa, M. *Willingness to Pay for Improved Electricity Supply Reliability in Zambia—A Survey of Urban Enterprises in Lusaka and Kitwe*; University of Cape Town: Cape Town, South Africa, 2018.

48. BPDB. *2018 Annual Report 2017-18, Bangladesh Power Development Board*; Government of Bangladesh: Dhaka, Bangladesh, 2018.
49. Ministry of Energy and Mines. Renewable Energy Data in Lao PDR. Available online: https://www.irena.org/eventdocs/Laos%20presentation.pdf (accessed on 16 May 2019).
50. Ethiopian Electric Power Corporation (EEPC), Power Sector Development - Powering Africa 2014. Available online: https://pubs.naruc.org/pub.cfm?id=537C14D4-2354-D714-511E-CB19B0D7EBD9 (accessed on 10 July 2019).
51. Johnson, O.; Muhoza, C.; Ogeya, M.; Lindstrom, A.; Granit, J.; Rosner, K. *Powering Africa-Unlocking opportunities for energy development in Southern Africa*; SEI Stockholm Enviroment Institute (SEI): Stockholm, Sweden, 2018.
52. MEWD. *Power System Development Master Plan for Zambia 2010-2030*; Ministry of Energy and Water Development (MEWD): Lusaka, Zambia, 2011.
53. Tripathi, A.K.; Iyer, P.V.R.; Kandpal, T.C.; Singh, K.K. Assessment of availability and costs of some agricultural residues used as feedstocks for biomass gasification and briquetting in India. *Energy Convers. Manag.* **1998**, *39*, 1611–1618. [CrossRef]
54. Purohit, P.; Tripathi, A.K.; Kandpal, T.C. Energetics of coal substitution by briquettes of agricultural residues. *Energy* **2006**, *31*, 1321–1331. [CrossRef]
55. Kumar, A.; Purohit, P.; Rana, S.; Kandpal, T.C. An approach to the estimation of the value of agricultural residues used as biofuels. *Biomass Bioenergy* **2002**, *22*, 195–203. [CrossRef]
56. IARI. *Crop Residues Management with Conservation Agriculture: Potential, Constraints and Policy Needs*; Indian Agricultural Research Institute (IARI): New Delhi, India, 2012.
57. Rasmussen, P.E.; Collins, H.P. Long-term impacts of tillage, fertilizer, and crop residue on soil organic matter in temperate semiarid regions. *Adv. Agron.* **1991**, *45*, 93–134.
58. Lal, R. The role of residues management in sustainable agricultural systems. *J. Sustain. Agric.* **1995**, *5*, 51–78. [CrossRef]
59. Vanlauwe, B.; Bationo, A.; Chianu, J.; Giller, K.E.; Merckx, R.; Mokwunye, U.; Ohiokpehai, O.; Pypers, P.; Tabo, R.; Shepherd, K.D.; et al. Integrated soil fertility management. *Outlook Agric.* **2010**, *39*, 17–24. [CrossRef]
60. FAO/UNEP. *A Decision Support Tool for Sustainable Bioenergy*; FAO: Rome, Italy; UNEP: Nairobi, Kenya, 2010; Available online: http://www.fao.org/3/am237e/am237e00.pdf (accessed on 20 September 2019).
61. Ackom, E.K. Sustainability standards for Canada's bioethanol industry. *Biofuels* **2010**, *1*, 237–241. [CrossRef]
62. Helwig, T.; Jannasch, R.; Samson, R.; DeMaio, A.; Caumartin, D. *Agricultural biomass residue inventories and conversion systems for energy production in Eastern Canada*; Resource Efficient Agricultural Production (REAP): Ste. Anne-de-Bellevue, QC, Canada, 2002.
63. Lal, R. Crop residues as soil amendments and feedstock for bioethanol production. *Waste Manag.* **2008**, *28*, 747–758. [CrossRef]
64. OECD/IEA. *Sustainable Production of Second-Generation Biofuels, Potential and Perspective in Major Economies and Developing Countries*; Information Paper; OECD: Paris, France; IEA: Paris, France, 2010.
65. Purohit, P. Economic potential of biomass gasification projects under clean development mechanism in India. *J. Clean. Prod.* **2009**, *17*, 181–193. [CrossRef]
66. Purohit, P.; Michaelowa, A. CDM potential of bagasse cogeneration in India. *Energy Policy* **2007**, *35*, 4779–4798. [CrossRef]
67. Bandyopadhyay, S.; Shyamsundar, P.; Baccini, A. Forests, biomass use and poverty in Malawi. *Ecol. Econ.* **2011**, *70*, 2461–2471. [CrossRef]
68. Purohit, P.; Fischer, G. *Second-Generation Biofuel Potential in India: Sustainability and Cost Considerations*; UNEP Riso Centre on Energy, Climate and Sustainable Development: Copenhagen, Denmark, 2014.
69. Gabisa, E.W.; Gheewala, S.H. Potential of bio-energy production in Ethiopia based on available biomass residues. *Biomass Bioenergy* **2018**, *111*, 77–87. [CrossRef]
70. Brecha, R.J. Threshold electricity consumption enables multiple sustainable development goals. *Sustainability* **2019**, *11*, 5047. [CrossRef]
71. Brecha, R. Electricity access threshold for meeting non-energy SDG targets. *Eur. J. Sustain. Dev.* **2019**, *8*, 90–100. [CrossRef]
72. Banerjee, S.G.; Bhatia, M.; Azuela, G.E.; Jaques, I.S.; Ashok, P.E.; Bushueva, I.; Angelou, N.; Inon, J.G. Global Tracking Framework. *Sustain. Energy World Bank* **2013**, *3*, 289.

73. The World Bank Group. *Beyond Connections: Energy Access Redefined*; World Bank: Washington, DC, USA, 2015; pp. 1–244.
74. Harmelink, M.; Beerepoot, M.; Puranasamriddhi, A. *Energy Access Projects and SDG Benefits Energy Access Projects and Assessment of their Contribution to the Sustainable Development Goals*; UNDP Discussion Paper; United Nations Development Programme: Bangkok, Thailand, 2018.
75. United Nations. *Household Size and Composition around the World*; United Nations: San Francisco, CA, USA, 2017.
76. Fritsche, U.R.; Eppler, U.; Fehrenbach, H.; Jürgen, G. *Linkages between the Sustainable Development Goals (SDGs) and the GBEP Sustainability Indicators for Bioenergy (GSI)*; Technical Paper for the GBEP Task Force on Sustainability. Global Bioenergy Partnership (GBEP), 2018. Available online: http://iinas.org/tl_files/iinas/downloads/bio/IINAS_IFEU_2018_Linkages_SDGs_and_GSIs.pdf (access on 24 April 2019).
77. Souza, G.M.; Ballester, M.V.R.; de Brito Cruz, C.H.; Chum, H.; Dale, B.; Dale, V.H.; Fernandes, E.C.M.; Foust, T.; Karp, A.; Lynd, L.; et al. The role of bioenergy in a climate-changing world. *Environ. Dev.* **2017**, *23*, 57–64. [CrossRef]
78. IRENA. *Bioenergy for Sustainable Development*; IRENA: Abu Dhabi, UAE, 2016.
79. RES4Africa Foundation Applying the water-energy-food nexus approach to catalyse transformational change in africa. 2019. Available online: https://www.res4africa.org/wp-content/uploads/2019/07/RES4Africa-Applying-the-water-energy-food-nexus-approach-to-catalyse-transformational-change-in-Africa.pdf (accessed on 25 August 2019).
80. WHO. The Equation Is Simple. 2016. Available online: https://www.who.int/sustainable-development/LR-HAP-27May2016.pdf?ua=1 (accessed on 20 October 2019).
81. Muyanja, D.; Allen, J.G.; Vallarino, J.; Valeri, L.; Kakuhikire, B.; Bangsberg, D.R.; Christiani, D.C.; Tsai, A.C.; Lai, P.S. Kerosene lighting contributes to household air pollution in rural Uganda. *Indoor Air* **2017**, *27*, 1022–1029. [CrossRef]
82. Lam, N.L.; Pachauri, S.; Purohit, P.; Nagai, Y.; Bates, M.N.; Cameron, C.; Smith, K.R. Kerosene subsidies for household lighting in India: What are the impacts? *Environ. Res. Lett.* **2016**, *11*, 044014. [CrossRef]
83. IPCC. *IPCC Guidelines for National Greenhouse Gas Inventories*; IPCC: Geneva, Switzerland, 2006.
84. Bhandari, R.; Stadler, I. Electrification using solar photovoltaic systems in Nepal. *Appl. Energy* **2011**, *88*, 458–465. [CrossRef]
85. GoN. *2016 Nepal's First NDC*; Government of Nepal (GoN): Kathmandu, Nepal, 2016; Volume 23, pp. 28–38.
86. Cardoso, J.; Silva, V.; Eusébio, D. Techno-economic analysis of a biomass gasification power plant dealing with forestry residues blends for electricity production in Portugal. *J. Clean. Prod.* **2019**, *212*, 741–753. [CrossRef]
87. Susanto, H.; Suria, T.; Pranolo, S.H. Economic analysis of biomass gasification for generating electricity in rural areas in Indonesia. In *IOP Conference Series: Materials Science and Engineering*; IOP Publishing: Bristol, UK, 2018.
88. Sobamowo, G.M.; Ojolo, S.J. Techno-economic analysis of biomass energy utilization through gasification technology for sustainable energy production and economic development in Nigeria. *J. Energy* **2018**, *2018*, 1–16. [CrossRef]
89. Bocci, E.; Sisinni, M.; Moneti, M.; Vecchione, L.; Di Carlo, A.; Villarini, M. State of art of small scale biomass gasification power systems: A review of the different typologies. *Energy Procedia* **2014**, *45*, 247–256. [CrossRef]
90. Abdelhady, S.; Borello, D.; Shaban, A. Techno-economic assessment of biomass power plant fed with rice straw: Sensitivity and parametric analysis of the performance and the LCOE. *Renew. Energy* **2018**, *115*, 1026–1034. [CrossRef]
91. Brigagão, G.V.; de Queiroz Fernandes Araújo, O.; de Medeiros, J.L.; Mikulcic, H.; Duic, N. A techno-economic analysis of thermochemical pathways for corncob-to-energy: Fast pyrolysis to bio-oil, gasification to methanol and combustion to electricity. *Fuel Process. Technol.* **2019**, *193*, 102–113. [CrossRef]

92. Luu, L.Q.; Halog, A. Rice husk based bioelectricity vs. coal-fired electricity: Life cycle sustainability assessment case study in Vietnam. *Procedia CIRP* **2016**, *40*, 73–78. [CrossRef]
93. Bhattacharyya, S.C. Viability of off-grid electricity supply using rice husk: A case study from South Asia. *Biomass Bioenergy* **2014**, *68*, 44–54. [CrossRef]

sustainability MDPI

Article

Evaluation of Water and Energy Nexus in Wami Ruvu River Basin, Tanzania

Mngereza Miraji [1,*], Xi Li [1], Jie Liu [1] and Chunmiao Zheng [1,2]

[1] Institute of Water Sciences, College of Engineering, Peking University, Beijing 100871, China; lixi09@pku.edu.cn (X.L.); jie.liu@pku.edu.cn (J.L.); zhengcm@sustc.edu.cn (C.Z.)

[2] School of Environmental Science and Engineering, Southern University of Science and Technology, Shenzhen 518055, China

* Correspondence: mngereza@gmail.com

Received: 9 May 2019; Accepted: 30 May 2019; Published: 2 June 2019

Abstract: In African nations, national and regional development targets for water and energy sectors seldom consider the nexus between the two, risking imbalances and inefficiencies in resource allocation and utilization. A typical example is the development and expansion of biofuel in the Wami Ruvu River Basin, Tanzania (WRB). Water Evaluation and Planning (WEAP) model was applied to the WRB to investigate the Water Energy Nexus (WEN), specifically, whether the development plan calling for biofuel expansion is a sound approach. Results show that WEN is much stronger in the biofuel irrigation consuming 69.3% and 61% of total biofuel's water and energy requirement, respectively. By 2035, the nexus continues to be stronger, consuming 54.5% and 49% of total biofuel's water and energy requirement, respectively, and thus first generation biofuels use much more resources in the growing than the refining process. An additional 768.2 million meter cubic of water and 413.4 million kWh of energy are needed for planned biofuel expansion, reallocating water to biofuels in water-scarce regions inherit related problems to other sectors such as increasing water use for the industry, agriculture, and energy sector by 67%, 45%, and 9%, respectively, which could further exacerbate stresses on water and energy supplies in the basin. Biofuel generation rely heavily on energy imports, as it consumes substantially more energy than it produces. Policies should promote the coordinated development of sustainable biofuel programs that are less water intensive with very low inputs of fossil fuels.

Keywords: water energy nexus; biofuel; Wami Ruvu Basin; Tanzania; Africa

---

1. Introduction

The sustainable management of water and energy resources faces enormous challenges globally in the upcoming decades due to the increasing complexity of dealing with them [1–3]. By 2050, global demand for energy will nearly double, while water and food demand is set to increase by over 50% [3]. Sustaining this upswing resource demand has driven the search for efficiency in water and energy sectors conjunctively, given competing needs for limited global resources in the era of climate change reality. The water–energy nexus has received global attention in recent years with a focus on how to balance their trade-offs [4], development agenda [5], interdependence [6], resources governance [7], and integrating one constraint into the other [8]. Proponents of the concept argue that by governing these resources in an integrated manner, problems such as resource scarcity, quality, and human wellbeing will be addressed in a more sustainable way [9]. The Water-Energy Nexus(WEN) captures the interdependency between the two resources, and focuses on the need for water in the Energy supply chain, and on the energy used to collect, clean, move, store, and dispose of water [10–13].

In the fast developing regions of the world, it is expected that the WEN approach applied to development plans may offer benefits such as increasing energy efficiency, decreasing water pollution, reducing costs of energy, and water delivery. However, WEN studies are few for the fast developing regions of Africa [14], despite well-known water scarcity issues facing 300 million people in sub-Saharan Africa [15–17]. The estimated reductions of about 20% in rainfall by 2080 from the current level of 650 mm/yr in the region is expected to exacerbate water, energy, and food insecurity if no action is taken [15,16], with indirect impacts on nutrition, health, sanitation, and water conflicts, among other social challenges.

Recent biofuel trends such as biofuel mandates in the EU [17] and USA [18] have heightened demand for bio-ethanol and biodiesel, thereby further contributing to the increased demand for land in sub-Saharan Africa and Southeast Asia [19]. With large biofuel land deals reported in Zambia, Sudan, Botswana, Namibia Mozambique, and Tanzania [20]. Sugarcane alone in Southern Africa occupies approximately 785,000 hectares (ha), of which more than 40% is irrigated. Several catchments are either closed or rapidly approaching closure [21,22], due to combined effects of resource over-allocation and poor rainfall to runoff conversion ratios sprinkled by high evaporation rates [23]. Several studies [19–21] pointed out the importance to take regional-specific characteristics and the multi-dimension approach into consideration when implementing biofuel policies [24].

Prior nexus studies often used WEAP modeling tool, with applications in dozens of basins across more than 170 countries worldwide [25]. Relevant for planning purpose is the model's ability to simulate climate change impacts on the water supply to different sectors [26–28], to optimize allocation priorities [29–31], to investigate supply-demand relation [32–34] and energy demands [35–37], and to conduct environmental assessments [38,39]. The model has enough flexibility to adopt different levels of data availability with a graphical interface, deals with distributed basin demand sites in large spatial scale with supply, and demand internally matched [40,41]. The model is considered to be the most suitable available methods for the nexus analysis at the basin level, due to its robust decision-making support and scenario generation features that govern the nexus [42]. Abdullahi et al. [26] applied the WEAP model and found that climate change will significantly reduce the runoff, and increases evapotranspiration in the Sokoto-Rima river basin in Nigeria; Höllermann et al. [27] revealed that the pressure on water resources will increase, decreasing inflows and groundwater recharge due to climate change aggravate the situation in the Ouémé–Bonou catchment in Benin; Kinoti et al. [31] used the model to balance the water requirements of competing users against the available water resources and found that the use of WEAP improved the complex system of demand-supply of the Ewaso Ng'iro river basin in Kenya. Johannsen et al. [33] used the model and found that climate change shows a significant decrease in available water resources and drastic reduction of irrigated agricultural area is the only solution to guarantee sustainable water use in the Drâa Basin in Morocco; Adgolign et al. [43] used the model and found that there will be 10.3%, reduction in the total annual flow and watersheds will have unmet demands towards the year 2050, but the impact of the existing and planned water resources development on delivery of instream flow requirements and downstream water availability is minimal in the Didessa Sub-Basin in Ethiopia. Despite the model being used to tested various water demand management options in various water-stressed basins in Africa [26,27,29,31,33,43,44], the WEN in these regions like the Tanzania's Wami Ruvu Basin remains unstudied and the influence of biofuel on water and energy availability and consumption have high levels of uncertainty. It is crucial to quantify the WEN and illustrate its complexities for effective operational and long-term water resource management, governance, and integrated water-energy resources management at the basin scale.

We evaluate WEN in the WRB of Tanzania where a development plan calls for 50% increase in biofuel production by 2025 [45,46]. The capital city of Tanzania, Dodoma as well as commercial and agriculture center of Dar es salaam and Morogoro respectively are located in WRB. In 2013, about 80% of the WRB population lives in urban areas, while 20% in rural areas—in complete opposition to the rest of Tanzania [47].

The aim of the study is to evaluate the WEN based on multi-sector water demands estimated under basin-developed scenarios and how uncoordinated biofuels agriculture impacts the basin's water resources. The study applied the WEAP model to simulate natural hydrological processes (e.g., evapotranspiration, runoff, and infiltration) and anthropogenic activities superimposed on the natural system to influence water resources and their allocation (i.e., consumptive and non-consumptive water demands) to enable evaluation of the water used for the entire Energy sector and energy used for the entire water use.

Following this introduction, the Materials and Methods part is found in Section 2, whereby, Study area, WEAP model development and input data, Scenario design, Country energy balance and the Water and energy for biofuel has been explained. Section 3 describes the results such as model performance, water and energy for biofuel in 2015, and water and energy for biofuel in 2035 were detailed as well. Section 4 includes discussion on resource security in the context of nexus, resource footprints management, and mitigation strategy in the nexus systems. Section 5 gives the general conclusion and some recommendations for future research. The conclusion shows the implications of the findings for the region and the world where multiscale biofuel expansion projects are being considered in various basins.

2. Materials and Methods

2.1. Study Area

Wami Ruvu Basin is located to the eastern part of Tanzania, which lies between Longitudes 35°30′00″ to 40°00′00″ E and Latitudes 05°00′00″ to 07°30′00″ S in the Eastern side of Tanzania, Figure 1. The area of the Wami Catchment is 43,742 km^2 and the total length of the Wami River is approximately 637 km. The area of the Ruvu Catchment is 11,789 km^2 and the total length of the Ruvu River is approximately 316 km. The basin area is the home of two estuaries, two national parks (Mikumi and Saadani), and two forest reserves. The Wami and Ruvu rivers are managed by the Wami-Ruvu Basin Water Office (WRBWO).

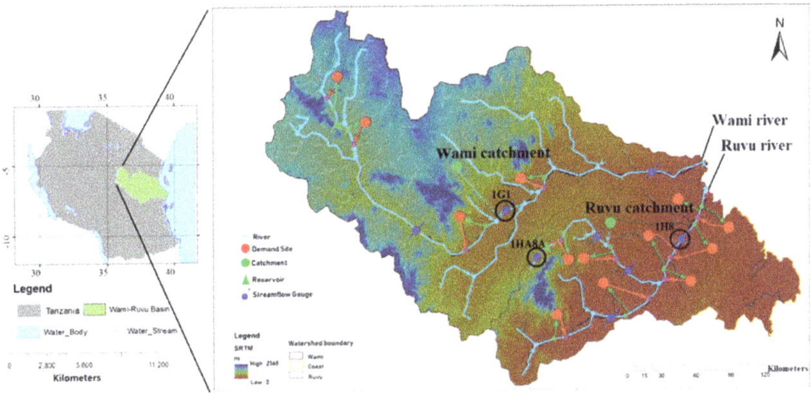

Figure 1. Map of the Wami-Ruvu Basin located in Eastern Tanzania.

2.2. Model Development and Input Data

As inputs to the model, climatic, hydrological, biophysical, and management data were collected from different sources and archives such as Wami Ruvu basin Water Atlas developed by Florida International University (FIU) [48], Integrated Water, Sanitation and Hygiene (iWASH) Program [49], and Global Water for Sustainability Program [50]. Sectorial water consumption (e.g., agriculture, domestic, industrial, energy, etc.), groundwater extraction, hydro-climate, and meteorological data (e.g., stream flows, temperature, precipitations, and rainfall) from various gauges were obtained from WRBWO's inventory.

The WEAP model was chosen because it is user-friendly, well documented, has enough flexibility to adopt different levels of data availability with a friendly graphical user interface, and has the advantage in dealing with distributed basin demand sites on a large spatial scale. The WEAP model was used to simulate natural hydrological processes (e.g., evapotranspiration, runoff, and infiltration) and anthropogenic activities such as agriculture and their water allocations between upstream and downstream users. Due to the high degree of interactions between water and energy related sectors, flow records at 1G1, 1HA8A, and 1H8, with 6%, 27%, and 26% missing data, respectively, were chosen to simulate demand and supply options. The gaps in the data were filled using the linear interpolation between the previous and next (non-missing) values.

The water system was characterized by: water demand sites, reservoir, flow gauging station, and river head flows. The study area boundaries were used to describe the spatial location of the water system. The demand site, reservoirs and catchments sources were represented with nodes. These nodes were linked to the Wami and Ruvu rivers through transmission links and return flow links. The demand is the product of total activity level and sectorial water user rates. The demand site nodes were created in the schematic view at relative positions, Figure 1. The demands were then named accordingly and demand priority set. Flow data for the year 2015 were used to create the current account (representing the system as it currently exists) in WEAP for scenario analysis.

After model construction and sectorial demand computation, the model performances were assessed using Nash and Sutcliffe coefficient, E, and Pearson's square method, R^2. The model interpolated (assign input time series for each WEAP catchment object for the climate data) inputs and created parameters (GDP growth, industrial growth, urbanization rate, population growth, and irrigation efficiency) for the entire simulation period. Monthly inflow data of 1955–1980 at the 1H8 (Ruvu river), 1G1 (Wami river) and at the 1HA8A (Ngerengere Catchment) gauging stations were used in simulating the flows for the 2016–2035 period. The model was run on a monthly time step (12 time steps per year), giving both $E > 50\%$ and $R^2 > 50\%$, which is the acceptable hydrological model performance. The current accounts year is set to be 2015, and then the model was run from the year 2016, beginning of projection to year 2035, and end of simulation by introducing three different set of scenarios.

2.3. Scenarios Design

An analysis of global datasets reveals a statistically significant relationship between water availability and per capita GDP [51,52]. The water-energy trade-offs to be tested using different water-centric scenarios, which is based on GDP growth rates.

For Scenario 1, the current GDP growth 6% and uncoordinated water management practices continue, neither water resources allocation nor priority setting between downstream and upstream water users are in place.

In Scenario 2, resource use management practices were in place, efficiency for water (drip irrigation, sprinkles, etc., 15–30%) and energy (pump efficiencies, 40–75%) were applied, then the biofuel demand was kept constant, only the population growth, GDP growth, industrial growth and urbanization rate were raised to 6%, 8%, 11%, and 85%, respectively, due to expected transformation to middle income countries.

In Scenario 3, energy intensifications (thermal + renewables energy deployments) accompanying with biofuel expansion were examined. The irrigated biofuel land was doubled whereby population growth, GDP growth, industrial growth, and urbanization rate were raised to 8%, 10%, 16%, and 90%, respectively, due to the expected socioeconomic development.

2.4. Country's Energy Balance

The most important measure in the energy balance of Tanzania is the total consumption of 5.7×10^9 kWh of electric energy per year. The balance is dominated by biomass-based fuels, constitutes 88% (5.02×10^9 kWh) of total energy consumption [53]. Commercial energy sources such as oil, gas, electricity, and coal, as well as non-biomass renewable energy, account for the remaining 12% (0.68×10^9 kWh), [54], Figure 2. The country has very low levels of electricity consumption per capita (100 kWh per person per year), far much less than the global average consumption of 2000 kWh per annum and average consumption in developing countries such as Sub-Saharan African of 552 kWh per annum [55].

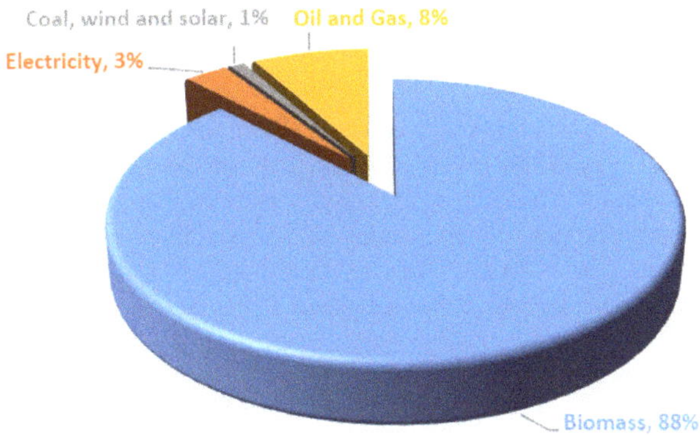

Figure 2. Tanzania's energy consumption pattern in 2015 [54,55].

2.5. Water and Energy for Biofuel

The estimates in Tables 1 and 2 are for three major estates in the Wami, Ruvu, and Coast catchment). The study consider water withdraw (from well or surface) as a measure of water use since they are continually metered by authority. Figure S1 in the Supplementary Materials show the relative position for all biofuel projects in the basin. First-generation biofuels are presently available biofuels produced using conventional technology, i.e., fermentation of carbohydrates into ethanol, and extracting and processing oil from oil crops into biodiesel [56]. Crops that are common used as biofuel feedstocks in Africa includes jatropha, oil palm, sugar cane, oil seeds, and sorghum [57]. Sugarcane is regarded as the main feedstock for the production of biofuel and considered in this study.

Energy intensity values for surface water and groundwater extraction in Tables 1 and 2 were adopted from Wakeel et al. [58] since there is no country specific data, while waste water and recycled water treatment based on widely used technologies in Africa such as lagoon ponds and trickling filter—as reported by Wang et al. [59] and Paul et al. [12].

3. Results

3.1. Model Performance

The model calibrated using the streamflow data monitored during the time period from 2005 to 2015, which contained both drier and wetter years. The sensitive analysis was used to determine the best value for some specific parameters. The model performances using the Nash-Sutcliffe Efficiency, E, and Coefficient of determination, R^2 [60], are calculated using Equations (1) and (2);

$$E = 1.0 - \frac{\sum_{i=1}^{N}\left(Q_{o_i} - Q_{s_i}\right)^2}{\sum_{i=1}^{N}\left(Q_{o_i} - \widetilde{Q}\right)^2} \tag{1}$$

$$R^2 = \frac{\sum_{i=1}^{N}\left(Q_{O_i} - \widetilde{Q}_{o_i}\right)^2 - \sum_{i=1}^{N}\left(Q_{s_i} - \widetilde{Q}_{s_i}\right)^2}{\sum_{i=1}^{N}\left(Q_{o_i} - \widetilde{Q}_{O_i}\right)^2} \tag{2}$$

where Q_{o_i} *is* the observed streamflow (m^3/s), Q_{s_i} *is* the simulated streamflow (m^3/s), $\widetilde{Q}$ *is* the average streamflow (m^3/s), and N is the total number of observations.

Equations (1) and (2) above indicated that performances at Wami River during calibration ($E = 0.69$, $R^2 = 0.74$) and validation ($E = 0.63$, $R^2 = 0.78$) and Ruvu River during calibration ($E = 0.80$, $R^2 = 0.86$) and validation ($E = 0.79$, $R^2 = 0.88$) were all satisfactory and acceptable model performances, (see Figures S2 and S3 in the Supplementary Materials).

3.2. Water and Energy for Biofuel in 2015

The agricultural sector accounts for 61% of total water consumed in the basin and biofuel irrigation constitutes 29% (251.9×10^6 m^3) of agricultural water use. Biofuel consumes 35% (104×10^6 m^3) of total water use in the energy sector to process bagasse in the energy (electricity and heat) production in the manufacture of ethanol. The industrial processes in the estates use 8% (10.5×10^6 m^3) of total water in the industry sector (Figure 3, Table 1, and Table S1). The WEN is much stronger in the biofuel irrigation sector consuming 69.3% and 61% of water and energy, respectively, thus biofuel use much more resources in the growing than the refining process (Table 1).

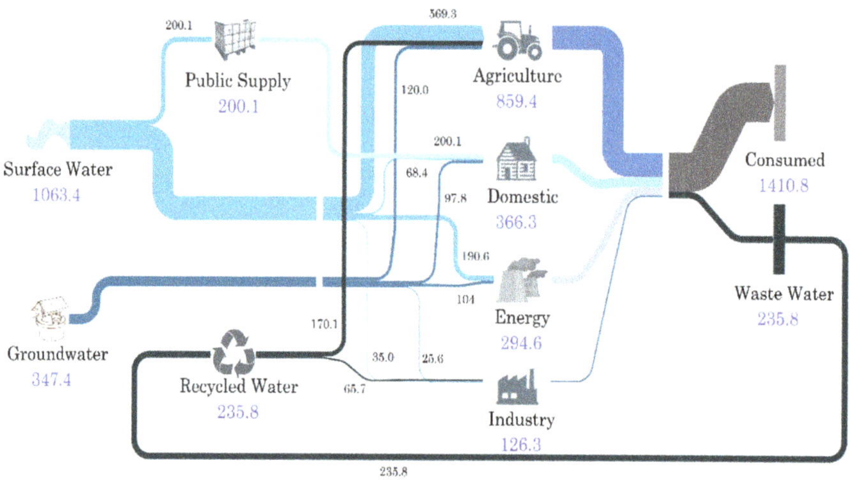

Figure 3. Sankey diagram of water flows from sources to destinations in 2015 (10^6 m^3). Sankey; diagram is made by eSankey (www.e-sankey.com).

Table 1. Water and energy needed for biofuel production in WRB (2015).

Items		Quantity (10^6 m^3/Year)	Energy Intensity (kWh/m^3)	Energy Consumed (10^6 kWh)	Data Sources
Surface water extraction	Rivers	247 (68%)	0.50	125.22 (61%)	WRB inventory, Wakeel et al. (2016)
	Reservoirs	4.91 (1.3%)	0.35		
Groundwater extraction		104 (29%)	0.75	78 (38%)	
Recycled water treatment		0.57 (0.1%)	0.48	0.27 (0.1%)	WRB inventory, Paul et al. (2018)
Waste water treatment		6.04 (1.6%)	0.28	1.69 (0.9%)	WRB inventory, Wang et al. (2016)
Total		362.52		205.18	

3.3. Water and Energy for Biofuel in 2035

3.3.1. With Biofuel Expansion

If the GDP turns to double digit (e.g., 10% growth) then the projected water needed for biofuel will be two times higher, additional 768.2 × 10^6 m^3 of water needed to accommodate the expansion. The energy needed for biofuel will grow by two-fold with additional 413.4 × 10^6 kWh needed for water extraction, recycling, and wastewater treatment (Table 2).

In this scenario biofuel irrigation share 36% (621.6 × 10^6 m^3) of the total agricultural water use and will consume 57% (312 × 10^6 m^3) of total water use in the energy sector to process bagasse into energy. Industrial water use for biofuel will increase from 8% (2015) to 12% (81 × 10^6 m^3) in 2035 (Figure 4, Table 2 and Table S1). The WEN continue to be stronger in the biofuel irrigation sector consuming 54.5% and 49% of water and energy, respectively, thus biofuel use much more resources in the growing than the refining process (Table 2).

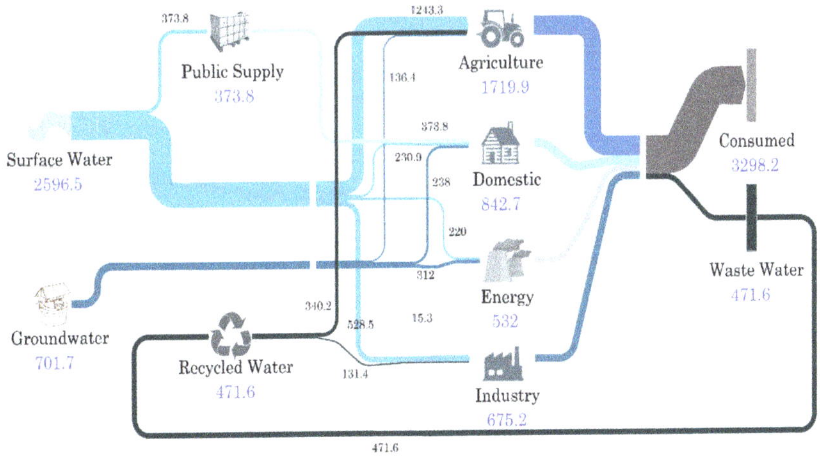

Figure 4. Sankey diagram of water flows from sources to destinations in 2035 (10^6 m^3). Sankey; diagram is made by eSankey (www.e-sankey.com).

Table 2. Water and energy needed for biofuel production in WRB (2035).

Items		Quantity (10^6 m^3/Year)	Energy Intensity (kWh/m^3)	Energy Consumed (10^6 kWh)	Data Sources
Surface water extraction	Rivers	570.5 (50%)	0.50	303.1 (49%)	WRB inventory, Wakeel et al. (2016)
	Reservoirs	51.1 (4.5%)	0.35		
Groundwater extraction		312 (27.5%)	0.75	234 (38%)	
Recycled water treatment		131.4 (12%)	0.48	63.1 (10%)	WRB inventory, Paul et al. (2018)
Waste water treatment		65.7 (6%)	0.28	18.4 (3%)	WRB inventory, Wang et al. (2016)
Total		1130.7		618.6	

3.3.2. Without Biofuel Expansion

Implementing resource use management practices without altering biofuel demand and resource efficiency for water (e.g., drip irrigation, sprinkles, etc.) being in place. With GDP growth 8%, the basin's water use for industry, agriculture, and energy will decline by 67%, 45%, and 9%, respectively (see Table S1 in the Supplementary Materials). Thus, water resources for industry, agriculture, and energy are becoming increasingly scarce in the basin as a result of increased competition with biofuel expansion.

Tables 1 and 2 shows that there is interdependence between water for biofuel and energy needed to support biofuels. Currently, biofuel consume 205.18 × 10^6 kWh, producing 138.1 × 10^6 kWh and will consume 618.6 × 10^6 kWh, to produce 290 × 10^6 kWh by 2035 (see Table S2 in the Supplementary Materials). This makes biofuel heavily rely on energy imports. Thus, it only provides a small benefit over fossil fuels in regards to greenhouse gases since they still require high amounts of energy to grow and process.

The current energy generation potential from excess bagasse in sugar mills is about 138.1 × 10^6 kWh per year that is 2.4% of the national electricity generation. By 2035, estimated potential from excess bagasse will be more than 290 × 10^6 kWh per year, which is 5.1% of the national electricity generation (see Table S2 in the Supplementary Materials). The biofuel is yet to replace the fossil fuel in the main grid system. The sugar and ethanol produced are only used for the domestic market.

4. Discussion

4.1. Resource Security in the Context of the Nexus

The impacts of biofuels on water resources can be mitigated by infrastructure expansion and allocation priority between upstream and downstream water users, which reduces surface water influence over total water abstraction and hence balance WEN trade-offs. Water use by the Energy sector is directly proportional to the water use for the biofuel irrigation that is related by efficient technologies in place. Despite the rapid sectorial growth, the nexus consumption trends are better saved under Scenario 2 (resource use efficiency) than under any other scenario. Mainly because high amounts of water savings are derived from technological advances and improved pumps efficiencies in the large irrigation areas that impacts a basin's water demand growth.

4.2. Resource Footprints Management

The current water and energy conservation measures in several basins in the developing regions such as Africa (e.g., Sokoto Rima River basin, Didessa Sub-basin, Mara River Basin, Niger River Basin, etc.) have largely focused water-efficient systems such as pressurized irrigations, which are more energy intensive [61]. Water and energy conservation can both be more effectively achieved by focusing on biofuel irrigation and its related sectors (e.g., water extraction) as well as improving pump efficiencies. Most water consumption occurs during the irrigation activities, which produce the biofuel feedstocks. Thus increasing global WEN productivity means decreasing the resource footprints for water, as well as mitigating their impacts on water resources.

4.3. Mitigation Strategies in the Nexus Systems

The demand surge of water and energy resources in basin areas can be mitigated by focusing on two major sectors, irrigation and surface water conveyance as both have major shares in most of the river basins in Nigeria, Ethiopia, Kenya, Morocco, Niger etc. The increase of agricultural withdrawal is due to rapid population, high urbanization and industrial demand. This has caused rising withdrawal to support livestock, crop-specific irrigation needs and energy (electrification) drive on the other hand. Meeting this rising demand could increase global water withdrawals in the energy sector by 20%, and water consumption in the sector by 85%, as indicated in the World Bank [2] and IRENA [3] reports. If water supply is reallocated to biofuels in water-scarce regions, this will pressure the other sectors that require it, which could inherit related problems. Thus, it is so important to optimize the freshwater efficiency of energy production as well as energy efficiency of water management, treatments, and distribution.

5. Conclusions

This study answers an important question on how to balance the resources tradeoffs by promoting coordinated development and why the water basins found in semi-arid regions such as West Africa, East Africa, Southern Africa, and South Asian countries should not apply freshwater resources to the production of Energy crops unless water energy nexus has been considered.

The nexus synergies and trade-offs should be balanced by the fact that, quantity of the water required for different energy production varies significantly according to the region, process, and technology of energy production, from rather negligible quantities of water used for solar and wind power generation to vast water use for biofuel irrigation, processing and production. For example, oil palms in Indonesia, Malaysia, and sugarcane in Brazil are grown in large quantities without irrigation due to the fact that the tropical lands they are grown on receive abundant rains, contrary to semi-arid regions such as WRB where sugar cane for ethanol is grown under irrigated conditions and water availability is a constraint.

First generation biofuels symbolize a step forward in energy independence and get-off fossil fuels for energy demands. These biofuels also support agricultural industries and rural communities through increased demand for crops production. That being said, first-generation biofuels, unlike second and third generation biofuels, they are extremely water intensive, requires lots of land to grow, compete with food crops over arable land in some parts of the world and still require high amounts of energy to grow, collect, and process.

Rapid increase of biofuel production could cause other environmental concerns in the basin such as water stress due to rapid population growth and abrupt increase in consumption. This study could be instructive to other regions facing similar problems. Highly urbanized, densely populated, loss of biodiversity, and freshwater scarcity regions share the concern that energy sector may scramble water resources with other sectors. Meanwhile, increase in energy production, water resources overexploitation as the global facts may have the trickle down effects on basin water budgets worldwide. Rational development policies and well-designed management practices are needed to ensure sustainable development of biofuels in this regard.

The analysis of nexus systems in the semi-arid basins is challenged by huge gaps in available data, inadequate water and energy utilities data that impedes careful nexus plans. These data are useful for a close examination of the nexus components and deserves increased attention in a world of growing scarcities to ensure optimal resource allocation to sustain living standards and preserve the environment.

Supplementary Materials: The following are available online at http://www.mdpi.com/2071-1050/11/11/3109/s1, Figure S1. Spatial distribution of biofuel projects; Figure S2. Calibration and validation hydrographs at (a) 1G1, Wami River (b) 1HA8A, Ruvu River; **Figure S3**. Calibration and validation scatter plots at (a) 1G1, Wami River (b) 1HA8A, Ruvu River; Table S1. Sectorial water use per scenario; Table S2. Biofuel generation in the Wami Ruvu Basin.

Author Contributions: Formal analysis, M.M.; Methodology, M.M. and X.L.; Resources, J.L.; Supervision, C.Z.; Visualization, X.L.; Writing–original draft, M.M.; Writing–review & editing, J.L.

Funding: This study was supported by the National Key Research and Development Program (No.2016YFE0102400) of the Ministry of Science and Technology of China.

Acknowledgments: The authors acknowledge the support provided by the WRBWO officials, Niqbar Ally and Pascal Qutaw, and Ally Mgelwa of the University of Dodoma. The authors thank the anonymous reviewers for their helpful comments.

Conflicts of Interest: The authors declare no conflict of interest.

Abbreviations

kWh, kilowatt hour; m^3, meter cubic.

References

1. IEA. Water for Energy: Is energy becoming a thirstier resource. In *World Energy Outlook*; IEA: Paris, France, 2012; pp. 1–33.
2. Rodriguez, D.J.; Delgado, A.; DeLaquil, P.; Sohns, A. *Thirsty Energy*; World Bank: Washington, DC, USA, 2013.
3. IRENA. *Renewable Energy in the Water, Energy and Food Nexus*; International Renewable Energy Agency: Abu Dhabi, UAE, 2015; pp. 1–125.
4. Zhang, X.; Vesselinov, V.V. Energy-water nexus: Balancing the tradeoffs between two-level decision makers. *Appl. Energy* **2016**, *183*, 77–87. [CrossRef]
5. Yumkella, K.K.; Yillia, P.T. Framing the Water-energy Nexus for the Post-2015 Development Agenda. *Aquat. Procedia* **2015**, *5*, 8–12. [CrossRef]
6. Tan, C.; Zhi, Q. The Energy-Water Nexus: A literature Review of the Dependence of Energy on Water. *Energy Procedia* **2016**, *88*, 277–284. [CrossRef]
7. Hagemann, N.; Kirschke, S. Key Issues of Interdisciplinary NEXUS Governance Analyses: Lessons Learned from Research on Integrated Water Resources Management. *Resources* **2017**, *6*, 9. [CrossRef]
8. DeNooyer, T.A.; Peschel, J.M.; Zhang, Z.; Stillwell, A.S. Integrating water resources and power generation: The energy-water nexus in Illinois. *Appl. Energy* **2016**, *162*, 363–371. [CrossRef]
9. Scanlon, B.R.; Ruddell, B.L.; Reed, P.M.; Hook, R.I.; Zheng, C.; Tidwell, V.C.; Siebert, S. The food-energy-water nexus: Transforming science for society. *Water Resour. Res.* **2017**, *53*, 3550–3556. [CrossRef]
10. Hoff, H. Understanding the Nexus. In *Background Paper for the Bonn2011 Nexus Conference*; Stockholm Environment Institute: Stockholm, Sweden, 2011; pp. 1–52.
11. Lee, M.; Keller, A.A.; Chiang, P.C.; Den, W.; Wang, H.; Hou, C.H.; Wu, J.; Wang, X.; Yan, J. Water-energy nexus for urban water systems: A comparative review on energy intensity and environmental impacts in relation to global water risks. *Appl. Energy* **2017**, *205*, 589–601. [CrossRef]
12. Paul, R.; Kenway, S.; Mukheibir, P. How scale and technology influence the energy intensity of water recycling systems—An Analytical Review. *J. Clean. Prod.* **2018**. [CrossRef]
13. Plappally, A.K.; Lienhard, V.J.H. Energy requirements for water production, treatment, end use, reclamation, and disposal. *Renew. Sustain. Energy Rev.* **2012**, *16*, 4818–4848. [CrossRef]
14. Endo, A.; Tsurita, I.; Burnett, K.; Orencio, P.M. A review of the current state of research on the water, energy, and food nexus. *J. Hydrol. Reg. Stud.* **2017**, *11*, 20–30. [CrossRef]
15. Thurlow, J.; Dalin, C.; Krueger, T.; Zhu, T.; Osborn, T.; Ringler, C.; Conway, D.; Dorling, S.; van Garderen, E.A.; Landman, W.; et al. Climate and southern Africa's water–energy–food nexus. *Nat. Clim. Chang.* **2015**, *5*, 837–846.
16. Mpandeli, S.; Naidoo, D.; Mabhaudhi, T.; Nhemachena, C.; Nhamo, L.; Liphadzi, S.; Hlahla, S.; Modi, A.T. Climate change adaptation through the water-energy-food nexus in Southern Africa. *Int. J. Environ. Res. Public Health* **2018**, *15*, 2306. [CrossRef] [PubMed]
17. Schuenemann, F.; Kerr, W.A. European union non-tariff barriers to imports of African biofuels. *Agrekon* **2019**, 1–19. [CrossRef]
18. Thompson, W.; Johansson, R.; Meyer, S.; Whistance, J. The US biofuel mandate as a substitute for carbon cap-and-trade. *Energy Policy* **2018**, *113*, 368–375. [CrossRef]
19. Chiarelli, D.D.; Davis, K.F.; Rulli, M.C.; D'Odorico, P. Climate change and large-scale land acquisitions in Africa: Quantifying the future impact on acquired water resources. *Adv. Water Resour.* **2016**, *94*, 231–237. [CrossRef]
20. Kiggundu, N.; Arhin, S.G.; Banadda, N.; Kabenge, I. Impacts of Biofuel Policies on Welfare and Food Security: Assessing the Socioeconomic and Environmental Trade-offs in Sub-Saharan Africa. *Int. J. Renew. Energy Res.* **2017**, *7*, 2162–2171.
21. Ngcobo, S.; Jewitt, G. Multiscale drivers of sugarcane expansion and impacts on water resources in Southern Africa. *Environ. Dev.* **2019**, *24*, 63–76. [CrossRef]
22. Serdeczny, O.; Adams, S.; Baarsch, F.; Coumou, D.; Robinson, A.; Hare, W.; Schaeffer, M.; Perrette, M.; Reinhardt, J. Climate change impacts in Sub-Saharan Africa: From physical changes to their social repercussions. *Reg. Environ. Chang.* **2017**, *17*, 1585–1600. [CrossRef]
23. Rodell, M.; Famiglietti, J.S.; Wiese, D.N.; Reager, J.T.; Beaudoing, H.K. Emerging trends in global freshwater availability. *Nature* **2018**, *557*, 651. [CrossRef]

24. McClain, M.E. Balancing water resources development and environmental sustainability in Africa: A review of recent research findings and applications. *Ambio* **2013**, *42*, 549–565. [CrossRef]
25. Stockholm Environment Institute (SEI). U.S.C. Water Evaluation and Planning System. In *Weap*; SEI: Boston, MA, USA, 2015; pp. 1–202.
26. Abdullahi, S.A.; Muhammad, M.M.; Adeogun, B.K.; Mohammed, I.U. Assessment of water availability in the Sokoto Rima River Basin. *Sci. Acad. Publ.* **2014**, *4*, 220–233.
27. Höllermann, B.; Giertz, S.; Diekkrüger, B. Benin 2025-Balancing Future Water Availability and Demand Using the WEAP "Water Evaluation and Planning" System. *Water Resour. Manag.* **2010**, *24*, 3591–3613. [CrossRef]
28. Santikayasa, I.P.; Babel, M.S.; Shrestha, S.; Jourdain, D.; Clemente, R.S. Evaluation of water use sustainability under future climate and irrigation management scenarios in Citarum River Basin, Indonesia. *Int. J. Sustain. Dev. World Ecol.* **2014**, *21*, 181–194. [CrossRef]
29. Mounir, Z.M.; Ma, C.M.; Amadou, I. Application of water evaluation and planning (WEAP): A model to assess future water demands in the Niger River (in Niger Republic). *Mod. Appl. Sci.* **2011**, *5*, 38–49. [CrossRef]
30. Mohd Firdaus Hum, N.N.; Abdul Talib, S. Modeling water supply and demand for effective water management allocation in Selangor. *J. Teknol.* **2016**, *78*, 15–20. [CrossRef]
31. Mutiga, J.K.; Mavengano, S.T.; Zhongbo, S.; Woldai, T.; Becht, R. Water Allocation as a Planning Tool to Minimise Water Use Conflicts in the Upper Ewaso Ng'iro North Basin, Kenya. *Water Resour. Manag.* **2010**, *24*, 3939–3959. [CrossRef]
32. Raskina, P.; Hansen, E.; Zhu, Z.; Stavisky, D. Simulation of water supply and demand in the Aral Sea region. *Water Int.* **1992**, *17*, 55–67. [CrossRef]
33. Johannsen, I.; Hengst, J.; Goll, A.; Höllermann, B.; Diekkrüger, B. Future of Water Supply and Demand in the Middle Drâa Valley, Morocco, under Climate and Land Use Change. *Water* **2016**, *8*, 313. [CrossRef]
34. Mourad, K.A.; Alshihabi, O. Assessment of future Syrian water resources supply and demand by the WEAP model. *Hydrol. Sci. J.* **2016**, *61*, 393–401. [CrossRef]
35. Ospina-Noreña, J.E.; Gay-García, C.; Conde, A.C.; Sánchez-Torres Esqueda, G. Water availability as a limiting factor and optimization of hydropower generation as an adaptation strategy to climate change in the Sinú-Caribe river basin. *Atmosfera* **2011**, *24*, 203–220.
36. Sattler, S.; MacKnick, J.; Yates, D.; Flores-Lopez, F.; Lopez, A.; Rogers, J. Linking electricity and water models to assess electricity choices at water-relevant scales. *Environ. Res. Lett.* **2012**, *7*, 045804. [CrossRef]
37. Flores-López, F.; Yates, D. A water system model for exploring electric energy alternatives in southeastern US basins. *Environ. Res. Lett.* **2013**, *8*, 035041. [CrossRef]
38. Flores-lópez, F.; Galaitsi, S.E.; Escobar, M.; Purkey, D. Modeling of Andean Páramo Ecosystems' Hydrological Response to Environmental Change. *Water* **2016**, *8*, 94. [CrossRef]
39. Gao, J.; Christensen, P.; Li, W. Application of the WEAP model in strategic environmental assessment: Experiences from a case study in an arid/semi-arid area in China. *J. Environ. Manag.* **2017**, *198*, 363–371. [CrossRef] [PubMed]
40. Stockholm Environment Institute. WEAP-TUTORIAL Water Evaluation and Planning System. In *Weap*; Stockholm Environment Institute: Boston, MA, USA, 2016; p. 286.
41. Purkey, D.; Galbraith, H.; Huber-Lee, A.; Sieber, J.; Yates, D. WEAP21—A Demand-, Priority-, and Preference-Driven Water Planning Model. *Water Int.* **2009**, *30*, 501–512.
42. Dai, J.; Wu, S.; Han, G.; Weinberg, J.; Xie, X.; Wu, X.; Song, X.; Jia, B.; Xue, W.; Yang, Q. Water-energy nexus: A review of methods and tools for macro-assessment. *Appl. Energy* **2017**, *210*, 393–408. [CrossRef]
43. Adgolign, T.B.; Rao, G.V.R.S.; Abbulu, Y. WEAP modeling of surface water resources allocation in Didessa Sub-Basin, West Ethiopia. *Sustain. Water Resour. Manag.* **2016**, *2*, 55–70. [CrossRef]
44. George, O.; Metobwa, M.; Mourad, K.A.; Ribbe, L. Water Demand Simulation Using WEAP 21: A Case Study of the Mara River Basin, Kenya. *Int. J. Nat. Resour. Ecol. Manag.* **2018**, *3*, 9–18.
45. Mshandete, A.M. *Biofuels in Tanzania: Small Scale Producers and Sustainable Environmental Management*; Land Rights Research & Resources Institute (HAKIARDHI): Dar es salaam, Tanzania, 2013.
46. Habib-Mintz, N. Biofuel investment in Tanzania: Omissions in implementation. *Energy Policy* **2010**, *38*, 3985–3997. [CrossRef]
47. Melesse, A.M.; Abtew, W.; Setegn, S.G. *Nile River Basin: Ecohydrological Challenges, Climate Change and Hydropolitics*; Springer International Publishing: Cham, Switzerland, 2013; pp. 1–718.

48. Saha, A.; Donoso, M.; Mohamed, A.; Masikini, R.; Abbott, V.; Anderson, E. *Water Atlas of the Wami/Ruvu Basin, Tanzania*; Florida International University: Miami, FL, USA, 2014.
49. GLOWS-FIU. *Socioeconomic Conditions and links to Freshwater Ecosystem Services of the Ruvu River Basin, Tanzania*; Global Water Sustainability Program: Morogoro, Tanznia, 2013; p. 38.
50. GLOWS-FIU. *Climate, Forest Cover, and Water Resources Vulnerability Wami/Ruvu Basin, Tanzania*; GLOWS-FIU: Morogoro, Tanzania, 2014.
51. Brown, C.; Lall, U. Water and economic development: The role of variability and a framework for resilience. *Nat. Resour. Forum* **2006**, *30*, 306–317. [CrossRef]
52. Brown, C.; Meeks, R.; Hunu, K.; Yu, W. Hydroclimate risk to economic growth in sub-Saharan Africa Hydroclimate risk to economic growth in sub-Saharan Africa. *Clim. Chang.* **2011**, *106*, 621–647. [CrossRef]
53. Kichonge, B.; John, G.R.; Mkilaha, I.S.N.; Hameer, S. Modelling of Future Energy Demand for Tanzania. *J. Energy Technol. Policy* **2014**, *4*, 16–32.
54. Kitonga, J.F.; Bonafe, E. *Tanzanian Energy Sector under the Universal Principles of under the Universal Principles of the Energy Charter*; The European Union: Brussel, Belgium, 2015.
55. Nyari, E.A.; Pogrebnaya, T.; Wilson, L. Energy sector and solar energy potential in Tanzania. *International Journal of Emerging Technologies and Engineering (IJETE)* **2015**, *2*, 143–149.
56. Gerbens-Leenes, W.; Hoekstra, A.Y.; van der Meer, T.H. The water footprint of bioenergy. *Proc. Natl. Acad. Sci. USA* **2009**, *106*, 10219–10223. [CrossRef] [PubMed]
57. Amigun, B.; Kaviti, J.; Stafford, W. Biofuels and sustainability in Africa. *Renew. Sustain. Energy Rev.* **2011**, *15*, 1360–1372. [CrossRef]
58. Wakeel, M.; Chen, B.; Hayat, T.; Alsaedi, A.; Ahmad, B. Energy consumption for water use cycles in different countries: A review. *Appl. Energy* **2016**, *178*, 868–885. [CrossRef]
59. Wang, H.; Yang, Y.; Keller, A.A.; Li, X.; Feng, S.; Dong, Y.N.; Li, F. Comparative analysis of energy intensity and carbon emissions in wastewater treatment in USA, Germany, China and South Africa. *Appl. Energy* **2016**, *184*, 873–881. [CrossRef]
60. Moriasi, D.N.; Arnold, J.G.; Van Liew, M.W.; Bingner, R.L.; Harmel, R.D.; Veith, T.L. Model evaluation guidelines for systematic quantification of accuracy in watershed simulations. *Trans. ASABE.* **2007**, *50*, 885–900. [CrossRef]
61. Siddiqi, A.; Fletcher, S. Energy Intensity of Water End-Uses. *Curr. Sustain. Energy Rep.* **2015**, *2*, 25–31. [CrossRef]

sustainability

MDPI

Article

Global Investment Failures and Transformations: A Review of Hyped *Jatropha* Spaces

Richmond Antwi-Bediako [1,*], Kei Otsuki [2,3], Annelies Zoomers [2,3] and Aklilu Amsalu [4]

1 International Development Studies, Department Human Geography and Planning, Faculty of Geosciences, Utrecht University, Vening Meinesz building 6.30, Princetonlaan 8a, 3584 CB Utrecht, The Netherlands
2 International Development Studies, Department Human Geography and Planning, Faculty of Geosciences, Utrecht University, Vening Meinesz building 6.12, Princetonlaan 8a, 3584 CB Utrecht, The Netherlands; k.otsuki@uu.nl (K.O.); e.b.zoomers@uu.nl (A.Z.)
3 Chair of International Development Studies & LANDac, Department Human Geography and Planning, Faculty of Geosciences, Utrecht University, Vening Meinesz building 6.06, Princetonlaan 8a, 3584 CB Utrecht, The Netherlands
4 Department of Geography & Environmental Studies, Addis Ababa University, Addis Ababa P.O. Box 150223, Ethiopia; aklilu.amsalu@aau.edu.et
* Correspondence: r.antwi-bediako@uu.nl

Received: 4 April 2019; Accepted: 11 June 2019; Published: 18 June 2019

Abstract: There was a heightened global interest in large-scale *Jatropha* cultivation for the past few decades, and this incited investment toward the crop in many developing countries. Many saw *Jatropha* as a green fuel that could possibly be an alternative to fossil fuel, which has adverse implications to deal with the impacts of climate change. However, *Jatropha* investments failed to meet global expectations, leading to unexpected social, environmental, and economic transformations in the investment spaces. This paper reviews and synthesizes the transformations and complexities in failed *Jatropha* spaces in six previous major *Jatropha* investment destinations across the world—Mexico, India, China, Ethiopia, Mozambique, and Ghana—by employing qualitative data analysis. The findings generally show that, in all of the countries studied, promoters of *Jatropha* investments, including the central government and private investors, subscribed to a "wait-and-see" approach with positive expectations. The review revealed that the intended goal of establishing global *Jatropha* investments to serve as an alternative source of fuel failed because of the unexpected complexities of the hype, which dwelled much on the deferment option of the "wait-and-see" approach for global *Jatropha* investments. Failure of the investments along with unmet expectations led to land-use changes from *Jatropha* to the cultivation of other crops (often food crops) or total land abandonment. Although we are not totally pessimistic about the economic and production viability of *Jatropha* as a biofuel feedstock, we emphasize the importance of paying considerable attention to other feedstocks that might have a better future as alternatives to fossil-based energy for the deployment of sustainable bioenergy. Furthermore, our findings provide meaningful justification for policy- and decision-makers in the development space to tacitly reflect and appraise new investment initiatives or interventions before endorsement.

Keywords: biofuel; *Jatropha*; investment; transformation; government; corporate; hype; bust

1. Introduction

Over the past decade, investment options toward large-scale *Jatropha* investments gained prominence around the globe. This was driven by the outrageous prices of oil in the international market, a desire to ensure energy security, and the quest to reduce the emissions of harmful gases leading to climate change [1]. According to Openshaw [2], *Jatropha* was favored by investors because the crop has many attributes, multiple uses, and great potential for meeting the global anticipation. For instance, the by-products of *Jatropha* can be transformed into a fuel source [3] and used for preparing soaps and candle wax [4]. The crop could also be used to impede soil erosion caused by rainwater and fencing, as well as to reclaim degraded lands [2,5]. Under this expectation, governments, multilateral organizations, and investors all entered into *Jatropha* investments [6].

According to Jingura et al. [7] and Tsegaye and Glantz [8], with growing global demand for climate-friendly transport fuels, countries and continents notably underwent large-scale land acquisition and investment for biofuels including African countries (Ghana, Ethiopia, Mozambique, etc.), Asia (China, India, etc.), and South America (Mexico, Brazil, etc.). Hence, governments in Africa, Asia, and South America partnered with countries, notably, Brazil and India, which are known to have considerable experience in biofuel development, to transfer technology into their biofuel sectors to invest in cultivating biofuel crops such as *Jatropha*, sugarcane, teak, and oil palm [8]. The interest of governments was further heightened by the acclaimed indications that *Jatropha* could improve local livelihoods, alleviate poverty, and kindle local development [9]. *Jatropha* investment took the center stage of agricultural investments in many countries south of the globe due to its immense contribution to socioeconomic development [10,11].

Historically, the origin/home of *Jatropha* is Mexico, and its knowledge and use were initiated by the Olmeca people of Mexico 5000 years ago [12], who planted *Jatropha* for medicinal purposes [13]. *Jatropha* then moved to Brazil and some portions of Central America. According to Henning [14], the Portuguese seafarers distributed *Jatropha* through the Cape Verde Islands to Africa and Asia. Openshaw [2] indicates that the potential of *Jatropha* as an oil-producing crop was long acknowledged in the form of demonstration farms in countries such as Nepal, Zimbabwe, Mali, and Brazil for the utilization of the oil. One of the first wide-scale productions of *Jatropha* was the Austrian–Nicaraguan project instituted in 1990. This project failed as investors saw it to be unprofitable [15]. Interest in energy security and climate change concerns re-ignited interest in *Jatropha* in the early 2000s. International funding bodies started to robustly support investors to invest in *Jatropha* [6,16] with the anticipation of ensuring sustainable energy security whilst having a positive impact on climate change mitigation [17]. *Jatropha* is argued to be a "marvelous crop" with huge potential for economic growth [2] and environmental management attributes [2,3] since it strives well even on less productive lands due to its lower requirements for water and non-competition for nutrients with other crops [18].

During the 2010s, there were burgeoning studies on the effects of large-scale *Jatropha* investments on environment and socioeconomic development [10,11,19,20]. The findings from these studies are mixed. On one hand, Bosch and Zeller [10] and the Food and Agriculture Organization (FAO) [11] showed that such investments enhanced employment opportunities, improved biodiversity, and increased revenue in the host communities. On the other hand, Schoneveld et al. [21] and Hughes et al. [20] provided evidence to suggest that the investments led to impoverishment, since the households' access to non-timber forest products, which are often considered as common-pool resources, is curtailed by such investments. The international Non-Governmental Organization (NGO), Action Aid [22], also disclosed that *Jatropha* investments created unintended consequences on total food production and the aggravation of poverty in investment-centered communities. These revelations, together with the quest to deal with energy poverty and climate change, motivated governments to support and allow large-scale *Jatropha* investments without sound preparation and arrangements. Preparations and arrangements in terms of studies on available and suitable lands, the legal incorporation of social and economic benefits, and conceptualized scientific studies were virtually absent [22]. Many of the countries did not have legal mechanisms to protect the interest, rights, welfare, and livelihoods of the rural population [23]. Even during the *Jatropha* investments, Skutsch et al. [24] and Birega [25] confirmed their doubt in *Jatropha* in terms of its potential to bolster Africa's rural development. The *Jatropha* hype without commensurate arrangements for its investments led to poor development outcomes including negative environmental implications, land alienation, rural livelihood loss, food insecurity, conflict, and high investment losses.

According to Timko et al. [26] and Hamenoo [27], outcomes of initial *Jatropha* investment were negative on local land tenure systems, food production, and livelihoods. The annual output of farmers declined due to their use of parcels of land that were smaller than the ones they had access to before the large-scale *Jatropha* plantations were established [28]. However, little was discussed about the relationship between the initial drivers of *Jatropha* investments, which were intended to do good, and the environmental and socioeconomic development outcomes of the investments and transformation patterns after the *Jatropha* investment failures, and how and why the *Jatropha* investments failed to generate the expected positive environmental and socioeconomic development outcomes, as well as what prospects the land had for future investments. This article presents a critical review of the published articles and literature on the *Jatropha* hype and bust, most of which were derived from the Sustainability journal special issue, dubbed "Global Jatropha Hype-Drivers and Consequences of the Boom and Bust of a Wonder Crop". This paper reviews and synthesizes the *Jatropha* investment transformation experiences and the drivers influencing *Jatropha* investment initiatives, government-driven initiatives, and private sector-driven initiatives impacting the production approaches of *Jatropha* investment, as well as how the transformation experienced influenced the environment, social, and economic systems in the six previous major *Jatropha* investment destinations of Mexico, India, China, Ethiopia, Mozambique, and Ghana.

Transformation and Investment Debates

Transformation theories are not entirely new in research dispensation, as they were long identified [29] in other fields, but are new in land grab and investment debates. The term transformation can elicit reactions ranging from organizational change and leadership to land-based investment intervention changes. According to Deming [30] and Daszko and Sheinberg [31], transformation is like a never-traveled journey and, as such, its destination is unknown, is tentative, and cannot be predicted, but welcomes new learning and actions based on new discoveries. Transformations in land spaces are a phenomenon of uncertainties engulfed with development decision-making at any given time [32]. Several models of decision-making under uncertainty were proposed including modern investment policy decisions [33]. In recent times, investment hypes like *Jatropha* were driven by uncertainties, randomness, and unpredictable events [34,35], and some uncertainties occurring in transformations are expected. Therefore, in the phenomenon of uncertainties in investment trends in such transformations, nothing happens out of nowhere, but outcomes are induced by cause–effect relationships with their attributed challenges. In some instances, unexpected events come on board by mere chance and/or an accident with or without recourse to modern investment, thereby adding to the risk element in modern investment. Even though the futuristic outcomes of transformations are unknown, predictions such as goal-setting are key in modern investment. According to Deming [30], predictions without information and knowledge constitute guessing, which leads to crisis-related risk. Muys et al. [36] support Deming's assertions and postulated that policy measures toward the adoption and promotion of new initiatives should hinge on multiple, interconnected, and viable data. Therefore, interconnected and viable data to predict policies and develop policy are key to promoting new initiatives. Lempert [37] revealed that a single policy implementation is usually insufficient to tackle a particular development problem and risk. Farazmand [29] highlights the more persuasive nature of risks associated with crisis in these contemporary times than before. The outcomes in modern large-scale investments include land grabbing, with significant changes in the transformation of property relationships and social relationships of production [38,39]. Kenny-Lazar [39] also postulated that modern large-scale investments may result from a variety of agrarian problems such as land degradation, larger firms out-pricing smallholder crops on the market, or the availability of more lucrative off-farm employment, which, before the investment, could not be predicted most probably due to the lack of an indigenous knowledge of the system. Keijzer and Lundsgaarde [40] did not consider whether unforeseen or unintended changes occurred, but instead focused on two lenses of why they occurred: (1) the occurrence of un-intendedness linked to human errors made in planning and implementing development interventions, and (2) the idea that reality is inherently complex to the extent that unintended effects would be unavoidable even when assuming the possibility of planning and implementing a perfect development intervention [41].

2. Methodology

To understand the outcomes of how *Jatropha* investments transformed global systems socio-economically and environmentally during and after the hype of *Jatropha*, our research reviewed the existing literature by employing qualitative data analysis. This was done through a case study design involving six countries that witnessed major *Jatropha* development (Mexico, India, China, Ethiopia, Mozambique, and Ghana) to describe the pattern of transformation in terms of undesired socio-economic and environmental outcomes. These countries were selected because they served as countries that witnessed major *Jatropha* development in a world where *Jatropha* investments were intensified through a mixed bag of government and corporate interventions. Again, after *Jatropha* was abandoned for almost a decade in these countries, no relevant studies were conducted to understand the transformational complexities of endorsing other crops. It, therefore, became very necessary to integrate the divergent unintended development consequences and transformational complexities of *Jatropha*. The qualitative data analysis helped provide insightful knowledge and understanding in a pragmatic manner [42,43] on investment transformation, which has very rare theoretical bedrocks and literature in the land grabbing debate. Secondary data collected through the review of literature from the Sustainability journal special issue "The Global Jatropha Hype-Drivers and Consequences of the Boom and Bust of a Wonder Crop" were reviewed to derive major themes as endorsed by Morse and Field [44] and Miles and Huberman [45] in tandem with each country by capturing the precise word that captured the key thought considered as "transformation in Mexico, transformation in India, transformation in China, transformation in Ethiopia, transformation in Mozambique, and transformation in Ghana". Based on the major themes, the findings were organized into a meaningful and logical cluster as suggested by Patton [46]. Furthermore, the analysis helped ensure that other relevant research findings were addressed in the discussion section of the study to inform the policy actions and academic thinking in a concretized manner.

3. Review of Country Cases: Transformations during the *Jatropha* Regime

A review of country-specific transformations for this article in the *Jatropha* space during the hype implies how approaches from investment sources transformed socio-economic and environmental systems during the era of *Jatropha*. Governments became the major driving force behind *Jatropha* investment during the *Jatropha* investment hype [47]. Government policies led to two main investment approaches and sources: (i) government-led investment approaches, and (ii) corporate-led investment approaches, which eventually defined the productions approach and investment sources, as well as the transformations emerging from these approaches (Figure 1).

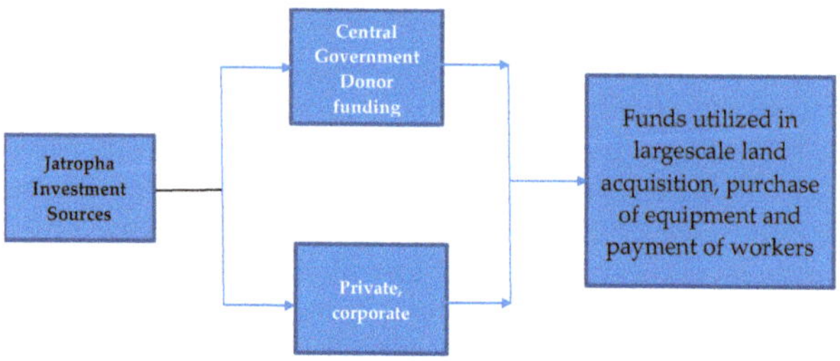

Figure 1. *Jatropha* investment sources and utilization.

3.1. Transformations in Mexico

In Mexico, the *Jatropha* investment sources and approaches were government-led. Mexico is known as the "home" of *Jatropha*, but the potential of *Jatropha* in producing biofuels was unknown until the early 1990s. The renewal in interest in *Jatropha* cultivation started in 2003, when its potential contributions to sustainable energy security drew global attention to invest in the crop [1,25]. As the "home" of *Jatropha*, the Mexican government enacted a national biofuel law, the 2008 Bioenergy Promotion and Development Act, as a quick response to (i) diversify Mexico's domestic energy production, (ii) commit to reducing greenhouse gases, (iii) rehabilitate and reforest degraded land through the ProArbol program of Comisón Nacional Forestal (CONAFOR), the National Forestry Commission, (iv) enhance rural employment, (v) stimulate rural development, and (vi) produce biofuel energy for local and global needs [24,47,48]. The primary focus of the law was to contribute to energy diversification and the sustainable development of the country's agriculture sector whilst making Mexico a prominent center of biofuel energy in the world, as well as contributing to the global goal of reducing greenhouse gases (GHGs) [48]. To sustainably move the 2008 act forward with biases, the Bioenergy Interministerial Strategy (2009–2012) was introduced with the aim of reducing political favoritism in other districts which lacked agronomical requirements to cultivate *Jatropha* [24]. The government had the task of obtaining additional energy sources for export [47,49], with the expectation of spearheading the future provision of additional energy sources to countries in the north. The drive for the biofuel policy was to enhance competitiveness and to "open the gate" for locals to utilize economic opportunities associated with biofuel investments. The government initiated *Jatropha* projects in the Chiapas state through the National Forestry Commission and instituted subsidies to achieve positive outcomes. Other states were Colima, Michoacán, Morelos, Oaxaca, Sinaloa, and Yucatan [50,51]. The government strategically encouraged *Jatropha* amidst other biofuel crops such as castor and oil palm, which were already known. *Jatropha* was tagged as the preferred feedstock for biofuel investment [52]. This preference was strengthened by the revelation that *Jatropha* could be successful on marginal lands [53] and the intention of the government to increase forest area [24].

The labor-intensive approach to investment was adopted to allow for the engagement of large-scale farmers to enhance their livelihoods through job creation and income provisions [24,54]. Both local and international private investors and agro-industries were also motivated to invest in *Jatropha* on a large scale, but the intention of the government was to limit their dominance [53]. The locals were, thus, made the key actors in *Jatropha* production. Farmers engaged in bean and maize production also resorted to the production of *Jatropha* [54,55]. Agro-businesses were involved in the production process, but the locals had a greater platform for *Jatropha* production.

Generally, *Jatropha* was invested in for more than three years in Mexico [25]; however, the investment outcomes were poor [55]. In states such as Michoacán, Veracruz, Chiapas, Quintana Roo, and Yucatan, the use of marginal land for the cultivation of *Jatropha* yielded poor returns [56]. It was expected that the use of marginal lands for *Jatropha* would not lead to negative implications on food production in the country [57], but bean and maize farmers moving into the *Jatropha* field contributed to falls in the production levels of these crops [24,53–55]. In addition, new diseases that caused harm to the production of other food crops affected *Jatropha* production [58], and *Jatropha* investment did not make any meaningful impacts on local livelihoods. Although new jobs were created for farmers [24], they could not obtain adequate and expected profits from their newly found jobs.

Valero et al. [55] highlighted that, in Chiapas, the motivation of the participants was to earn enough income through the sale of the *Jatropha* seeds. This motivation behind the farmers in Chiapas was similar to the Mexicans in Quintana Roo, who were selling the seeds for biofuel production and export for revenue [49,59,60]. With frustration growing from the investment, some local individuals and agro-industries opted out of the production of *Jatropha*, and, in Chiapas State, authorities forewent the *Jatropha* project. Institutions and refinery centers that were previously created were closed down [58] as the production could not lead to a substantial reduction in poverty as proclaimed in many studies, but instead deepened the levels of poverty of its actors [24].

The environmental implications of *Jatropha* were generally a mixture of positive and negative outcomes [24]. Regarding positive environmental implications, the *Jatropha* project led to afforestation/reforestation in Yucatan; for instance, Skutsch et al. [24] observed that about 2000 hectares and 350 hectares of lands were used in two different commercial ranches to expand the forest. However, in Michoacán, the *Jatropha* farmers adopted a shifting cultivation approach, which saw a significant number of farmers clearing the secondary forest on a rotational basis for *Jatropha* plantations. Mexico's government, corporate investors, and individual local farmers felt disappointed in the outcome of *Jatropha* in terms of spearheading economic transformation and meeting energy needs [55]. As a result, the National Biofuel Policy shifted its ultimate attention from *Jatropha* to focus on other biofuel crop research projects and development (Mexican Bioenergy Network as in GAIN [61]). Investors withdrew their investments and some *Jatropha* lands currently lie idle, whilst others are being used for food crop production by previous *Jatropha* farmers [58].

3.2. *Transformations in India*

In India, *Jatropha* investment sources and approaches were government-led initiatives through the Indian Government's declaration of National Mission on Biofuels in 2003 [62]. The global interest in biofuel investments aroused the awareness of the Indian Government to invest in *Jatropha*, [63] as in the case of Mexico. The government saw *Jatropha* cultivation as a means to improve soil fertility, contribute to the reduction of soil erosion, help in the rehabilitation of lands through greening, and create jobs for local members [63]. Even though several biofuel crops such as palm trees, maize, and sugar cane had higher attention from the government, *Jatropha* was ranked as the most preferred biofuel crop based on the following reasons: the easy cultivation of *Jatropha*; the faster growth and hardy nature of *Jatropha*; the seeds of *Jatropha* are easy to collect as they are always ready to be plucked before the rainy season; *Jatropha* plants are not very tall, but cannot be browsed by animals; the crop is rich in nitrogen; and the seed cake is a source of plant nutrients. Based on the attributes of *Jatropha*, the government realized that the cultivation of the crop could help the country achieve the 20% blending target [62,64]. About 11.2 million hectares of land was needed for the cultivation of *Jatropha* in the country [64].

Although the major source of investment was expected from the government through its policy which revealed its total commitment in supporting local actors in *Jatropha* enterprises, the implementation saw corporate local actors and farmers using their financial resources for the cultivation of *Jatropha* [64].The government policy paid credence to massive local involvement, as well as the engagement of private and corporate investment through minimum support price mechanizations, which were proposed in the Biofuel Policy of India [65]. To ensure active and massive local participation in the production of *Jatropha*, the government proposed a special support package for the locals who wanted to be involved in *Jatropha* cultivation. This proposed support was quite different from Mexico's case, as the government never promised to directly incentivize local *Jatropha* farmers. In the Andhra Pradesh State of India, the government promoted *Jatropha* through incentives in the form of free seedlings provided for farmers, a 90% subsidy for the installation of drip irrigation systems, and free land preparation under the Food for Work and Assigned Land Development program [62]. Poor farmers who were assigned lands owned by the government were to be provided with full financial assistance by the government with 40% secured as a bank loan for the cultivation of *Jatropha*. In the state of Tamil Nadu, companies and non-governmental organizations were encouraged to partake in the cultivation of *Jatropha*. Thus, India's *Jatropha* cultivation was made up of massive local participation, as well as a few other private entities that were mainly local companies and organizations, due to the government's enticing proposals in its policy.

Axlesson and Frazen [62] found that, in certain parts of India, *Jatropha* was cultivated on productive lands, but still failed. Therefore, it became very complex for stakeholders to explain why such production could not yield enough output. Studies also acknowledged that *Jatropha* has positive environmental outcomes by reducing the emissions of greenhouse gases [3]. *Jatropha* production was perceived to have an insignificant contribution to climate change [66] since the number of plants per hectare of Indian farms was unlikely to have an important effect on the uptake of carbon [64]. This revelation supports the position of Prueksakorn and Gheewala [67] that the cultivation of *Jatropha* has insignificant adverse impacts on climate change. Indian engagement in *Jatropha* production led to intense pressure on the limited water resources, which was due to the demand of water for irrigation purposes. Although it is perceived that *Jatropha* is a drought-resistant crop, farmers realized that water was needed for positive yields. The government developed a financial plan with the quest to provide financial incentives, as well as input to support the production of *Jatropha* by the locals [62]. The government's failure to deliver its promises affected *Jatropha* cultivation, and the locals were forced to use their limited financial resources for the production process, to a point where they could no longer finance their investments. As a result, they withdrew their efforts. Additionally, the farmers also had very limited knowledge concerning the cultural practices and maintenance of *Jatropha* plantations. Extension officers failed to periodically visit to educate the farmers on the maintenance practices needed for successful outputs, which also contributed to the disappointing yields. The national biofuel policy failed to provide clear marketing plans for *Jatropha* products such as biodiesel, and there were no communication plans and strategies drawn and implemented to ensure the flow of marketing information between the farmers and biodiesel producers to ensure sustained demand. This led to the absence of a market for *Jatropha* products and made the *Jatropha* business unattractive for the local farmers to engage in. Thus, instead of *Jatropha* investment contributing to poverty alleviation, it further exacerbated it [62].

Axlesson and Frazen [62] further elaborated that poor cultural practices due to lack of requisite technical management support from agricultural extension officers paved the way for pests and diseases to attack *Jatropha* crops. In Tamil Nadu State, for instance, the pest attacks were generally mild and could have been tackled with the use of pesticides; however, in a few instances, the attacks were more severe than expected and this adversely affected the growth of the *Jatropha* plants, leading to their forced removal. The common pest identified was the mealy bug, which destroyed several crops. *Jatropha* production saw the creation of jobs, but this did not really contribute to advancements in local livelihoods. Issues on land alienation were virtually absent as the government largely promoted private land use for the cultivation of *Jatropha* and incentivized local farmers to use their own land for the cultivation of *Jatropha*, which were government lands in some cases [62]. The government, investors, and local members halted *Jatropha* investment, and some *Jatropha* lands are now occupied by local members, whilst others are yet to be approached. The government subsequently lifted its focus on *Jatropha* to a wide array of potential feedstocks including palm trees and sugarcane, with the hope of producing biofuels at a cost-convenient level [68].

3.3. Transformations in China

Even though China has a centrally controlled economic system, *Jatropha* investment sources and approaches leant toward a corporate-led production scheme with close monitoring by the government. According to Li et al. [69], in 2005, the government of China took the lead in developing a policy environment to allow for biofuel cultivation, one being the "Renewable Energy Industry and Development Supervision" policy. This policy paid credence to supporting biofuel plantations through the selection of a better variety of crops. The government's initial attention was geared toward technological research and demonstration projects. Therefore, the government passed the National Forest Construction Plan in 2006, which aimed to provide about 13 million hectares of energy forest by 2020, as well as supply raw materials for about six million tons of biodiesel. In 2007, several plans of actions were developed, where significant among them was the "Medium–Long-Term Development Plan on Renewable Energy", which paid attention to (i) the production of biofuel plants; (ii) the establishment of *Jatropha*-breeding biodiesel experiments in some selected provinces; (iii) an increase in the yearly target for the utilization of biofuel with an expected increment of 1.8 million liters by 2020; and (iv) the management instruction on financial subsidies supporting crops for energy. In 2011, the Chinese Biofuel Policy on *Biodiesel Fuel Blend* had the aim of blending 2–5% biodiesel with 95–98%

diesel. With these government policies, several governmental bodies were engaged, including the Chinese Ministry of Finance and the Ministry of Agriculture, to promote the development of *Jatropha* production across the country. In Sichuan Province, about 14,667 hectares of land was acquired for *Jatropha* cultivation in 2007. Other provinces were Guanjxi and Yunnan. Concerned ministries came together to coordinate how best the country could engage in the investment of *Jatropha* to meet their energy demands and for export [70]. Furthermore, there were plans to increase the cultivation of *Jatropha* [70]. The southwestern part of China was seen as a suitable avenue for *Jatropha* due to the numerous hectares of unproductive lands for cultivation [71].

The investment sources of *Jatropha* were from investors in collaboration with the government. The government mainly played a supportive role, but the actual funds for *Jatropha* cultivation came from the investors. The government had an interest in enhancing well-being and dealing with poverty amongst the rural population; hence, the plan was to allow the locals to engage in the production of *Jatropha*. Corporate bodies could employ locals and train them in the production process, thus ensuring higher yields whilst enhancing the living conditions of the people [69]. Corporate bodies, including international investors and individuals, were engaged in *Jatropha* production. Two provinces, Sichuan and Guanjxi, were the two predominant centers for the cultivation of *Jatropha* in China. Corporate bodies employed community members to work as field workers/laborers for the production process; thus, the production approach was to make *Jatropha* production labor-intensive to allow for employment generation and advancement in local livelihoods.

The Chinese government was unable to achieve its expectation of meeting the growing energy needs of the local people using *Jatropha*. According to Li et al. [69], the poor government support, the limited demand for biofuels produced from *Jatropha*, and the severe frost led to the withdrawal of corporate investment, which contributed to the poor outcomes. The poor investment outcomes made the corporate entities leave without compensating the farmers for their services. In Guangxi, for instance, a private corporate company known as the Guangxi Zhilian Renewable Energy Company was involved in the *Jatropha* investment. The company employed farmers using a contract farming scheme. The company withdrew its investment within a short period of operation because the company needed subsidies from the central government, which were not forthcoming. During the periods of operation, the companies paid the workers through an investor–government coordinated system. This, however, failed and, as a result, the workers' livelihoods did not see massive improvements [69]. It is purported that the heightened interest in biofuel investment led to dynamics in land-use rights and land tenure security of the local participants [23,72]. However, in China, the *Jatropha* investment did not alter land-use rights and tenure security [69]. This was because the lands used were government-acquired land, and, as such, community members were not affected. However, *Jatropha* production and processing expenses escalated its market value higher than fossil products, thereby affecting demand and profitability. This blocked the investors' interest, as the government failed to grant them subsidies, leading to the *Jatropha* bust [69].

3.4. Transformations in Ethiopia

Even though Ethiopia has a centralized controlled economic system, the *Jatropha* investment sources and approaches leant toward a corporate-led production scheme with close monitoring by the government using the "Open-Door Policy" [73–76]. The *Jatropha* era saw Ethiopia as an attractive destination for advanced countries to undertake large-scale investment in biofuel production, particularly in *Jatropha*. Faced with global pressures from foreign-based investors, the government was concerned about how the country could take advantage of the *Jatropha* hype to create a greener economy and produce energy to meet the local needs. The government developed the Biofuel Strategy to ensure that its aims were achieved [73]. The strategy paid attention, first and foremost, to ensuring a green economy using *Jatropha* biofuel production, as well as improving agriculture, ensuring natural resource management, and alleviating poverty in the country [74]. *Jatropha* investment was fueled by two main forces: the government's quest to secure energy through local production and global forces due to the increase in demand for biofuels on the international market [75]. In 2006, a biofuel company known as Sun Biofuels in the United Kingdom (UK) started its *Jatropha* investment in the country. By 2010, Ethiopia had about 83 licensed foreign companies to invest in biofuel production [76]. Bossio et al. [77] revealed that foreign direct investments in *Jatropha* amounted to 50% of the total foreign investment in the country in 2011. A land deal matrix in 2012 indicated that about 1,360,670 hectares of land was released for biofuel projects with more than 700,000 hectares being used for *Jatropha* investment [78]. The global interest and pressure on Ethiopia influenced the government to make two main policy changes. The first was the government's introduction of the Ethanol Blending Policy, and the second was the changes made to the agriculture development and taxation policies, making the country an attractive center for foreign biofuel investors. This policy, from the perspective of Rahmato [79], was known as the "Open-Door Policy". In 2007, the government further developed a "desperate" policy dubbed the Biofuels Development and Utilization Strategy [75]. The policy spelt out the approach to enhance biofuel production and use within the country, and marginal lands of about 23.2 million hectares were earmarked for commercial *Jatropha* production [80].

The major source of investment was from foreign investors. The investors acquired large land sizes for *Jatropha* enterprises, and the government played a facilitating and supporting role in the entire *Jatropha* experience [75,76]. The flexible and open-door policy of the Ethiopian government saw the influx of foreign investors acquiring large hectares of land for *Jatropha*. The corporate investors employed the local people to work on the plantations on a paid scheme. Some of the locals were permanent workers whilst others were temporal. The production approach involved the use of both labor and machines on the plantation farms [81].

The government was very much interested in ecological integrity, which explains why the Climate-Resilient Green Economy Strategy was developed. The *Jatropha* investment did not make any significant contribution to a greener economy in Ethiopia after the bust, but Portner et al. [73] indicated that the *Jatropha* production had great potential of reducing soil erosion and retaining water for agriculture production in Ethiopia. In terms of livelihood benefits, *Jatropha* failed, as its production did not make any substantial impact on positive livelihood transformations. *Jatropha* could not be used to alleviate poverty, which explains why Portner et al. [73] elaborated that, for farmers to have economically benefited from *Jatropha*, there had to have been measures placed on their workloads, the local processing of seeds, training programs, and market interconnectedness for *Jatropha*. The investment saw the local farmers' land alienation, which served as their economic asset. Power holders and investors did not have respect for local land rights, and, as such, individuals whose farms were on the investors' acquired land were alienated, and this worsened the poverty in the project-affected communities [7]. Some local communities did not support the *Jatropha* project and/or were promised developmental co-benefits such as basic infrastructure and employment opportunities. The findings confirmed the assertion of the UN-Energy[1] [82] that biofuel production such as *Jatropha* plantations did not entirely benefit the poor farmers; rather, they put them into "deeper" poverty. This explains why Openshaw [2] indicated his doubt in *Jatropha* as a "poor man's crop". In the direction of the implications of the *Jatropha* investment on food production, the study by von Maltilz et al. [81] in Niqel, a *Jatropha* investment destination in Ethiopia, revealed changes in food availability and accessibility due to income received from the involvement of the locals in the *Jatropha* production process. According to their study, the changes were mixed. In some areas, food production was negatively affected, as farmers using grazing and farmlands for rotational farming were alienated. The widespread outcome of *Jatropha* was negative, as investment proved impracticable to continue. Hence, they abandoned their acquired lands, thus allowing local farmers to utilize some areas of the *Jatropha* land for food crops [81].

3.5. Transformations in Mozambique

Even though Mozambique has a centralized economic system, *Jatropha* investment sources and approaches leant toward a corporate-led production scheme with close monitoring by the government.

Mozambique is amongst the countries where *Jatropha* investment took place. As indicated by Mataveia [83], the country is highly dependent on the international market with about 700,000 cubic meters in the annual consumption of petroleum products. During the phase of global forces for the widespread production of biofuel crops such as *Jatropha*, Mozambique also needed to enhance its energy security and reduce its over-dependence on international oil products. According to Schut et al., [57] the National Biofuel Policy and Strategy (NBPS) for the country was, therefore, developed in 2009 based on the following motivations: the unstable and volatile nature of oil prices in the world market; biofuel as an alternative energy product to reduce the dependency on fossil fuels; and the reduction in the emission of greenhouse gases (GHGs), as well as encouraging the use of safe and clean energy in the country [83]. The policy further emphasized poverty reduction and focused on encouraging private sector participation through collaboration and networking with the government, ensuring cross-sectoral coordination in the country through strengthening inter-institutional collaborations and frameworks comprising ministries, departments and agencies, tertiary institutions (notably, universities), financial bodies, non-governmental organizations, and civil groups in biofuel development. The country further deployed the Kyoto Protocol Mechanisms and other international instruments with the quest to speed up the utilization of green fuels and to make an impact on the environment through the reduction of greenhouse gases (GHS) [83].

The major sources of investment were from the central government's coffers and corporate funding. The government established a network with interested private investors both locally and internationally to ensure effective investment in *Jatropha* enterprises in the country [81]. The government allowed the private sector to take dominance in *Jatropha* cultivation, which allowed the locals to obtain employment in the *Jatropha* companies. The private companies partnered with the government during the production processes, and the local people participated in *Jatropha* cultivation on individual farms. The government had the intention to collaborate with the tertiary institutions to provide theoretical footprints to support the investment [81], but this was unsuccessful. During the production processes, the government mainly focused on small holders and communities by supporting them to cultivate *Jatropha*. Key production actors were foreign investors, local private entities, and individual farmers, who either worked for the investors or planted *Jatropha* on their own.

The government of Mozambique showed great commitment to *Jatropha*. Even with poor initial production outcomes in 2007, the government encouraged investors to continue with the *Jatropha* project by providing tax incentives [84]. Regardless of such incentives, *Jatropha* failed as a biofuel crop. *Jatropha* business was unprofitable, and the farmers withdrew their investment and efforts [85]. Instead of *Jatropha* becoming a safety net, it became a huge risk venture for the local people. Mozambique was noted to have used large hectares of land for *Jatropha* cultivation [86–88], and, during the investment, issues on land rights and tenure security became a critical development challenge in the country. Aggrieved community members closer to the investment sites did not support the project since their lands were taken from them without their due consent. Mozambique also had issues with the influx of diseases and pests [84], where the adverse effects of these diseases and pests are yet to be made known. Generally, van Eijck et al. [89] noted that *Jatropha* production in Mozambique received mixed results on the farmers' food production. This is because, whilst some community members stated that *Jatropha* production adversely affected their food production, others witnessed improvements in meeting their food needs. Schut et al. [57] realized that the outputs and income obtained from *Jatropha* production were lower compared to other crops; thus, it was better to utilize available land space to produce cash or food crops instead of *Jatropha*. In terms of job creation, Bos et al. [85] postulated that many of the locals were employed by investors during the *Jatropha* experience. The study by Romijn et al. [90] confirmed that more than 500 permanent jobs were created due to the investment decision in favor of *Jatropha*.

3.6. Transformations in Ghana

Jatropha investment sources and approaches in Ghana were similar to those in Mozambique and Ethiopia, where there was a corporate-led production scheme with some monitoring mechanisms and interest from the government. The *Jatropha* investment in Ghana was started by a corporate entity known as Annuanom Industrial Project Limited [91]. In 2003, this entity called on the government to consider *Jatropha* investment as an innovative avenue to ensure local development transformations [92]. Through background studies, the government was convinced to commit its resources to *Jatropha* investments. According to Ahmed et al. [93], a National *Jatropha* Project Planning Committee was set up to assist in establishing *Jatropha* plantations in the country. The committee recommended pilot *Jatropha* plantations in 53 districts over a period of five to six years, on unproductive and dormant lands [94,95]. The government was expected to lead the *Jatropha* investment initiative by engaging local farmers to transform their livelihoods positively [92]. The committee suggested that these farmers should be trained extensively by Ghana's agriculture ministry [96] to ensure that they gained knowledge on the agronomy practices of *Jatropha*. A market strategy specified that the government should purchase the outputs of the farmers as the biofuel policy-mandated government-owned vehicles were to run on biodiesel on a minimum of B20 [96].

Technoserve [97] unveiled that the Ghanaian government had two biofuel crops to choose from: oil palm and *Jatropha*. The government made its choice in favor of *Jatropha* and decided to pump funds toward its cultivation. The Ghana Energy Commission [98] confirmed that the government was dedicated to *Jatropha* investment because of the huge importation cost of crude oil in 2004, which ranged from around United States dollars (USD) 516.8 million to USD 816.1 million. Unfortunately, the intended *Jatropha* project to be implemented by the government in the selected districts came to a halt when Ghana discovered crude oil in 2007. Therefore, both foreign and local investors were granted the opportunity to invest in *Jatropha* [99]. These investors acquired large stretches of land for *Jatropha* plantations [21,91] through negotiations with the chiefs [100]. Energy-inclined civil society groups such as the Kumasi Institute of Technology, Energy, and Environment (KITE-Ghana), the Gratis Foundation, and New Energy-Ghana also engaged in *Jatropha* investment, but on a smaller scale with funding support from United Nations Development Programme (UNDP). These organizations had an interest in using *Jatropha* to have a positive impact on communities [101,102].

Jatropha investment in Ghana generally unfolded transformations as unsatisfactory development outcomes for both investors and local communities. In a study conducted by Timko et al. [26], several outcomes were identified in some selected *Jatropha* investment sites in Ghana. Firstly, an average of 55.5 acres of land utilized by farmers were taken over by investors for large-scale *Jatropha* production. The takeover of farmers' land was never expected, as the motivation for the push for *Jatropha* was that *Jatropha* could be successful on marginal lands (supported by Reference [53]). Hence, there was the stance that local farmers would not be alienated from their farmlands [103]. Most of the large-scale investments were featured on productive lands, leading to the alienation of farmers. The second revelation by Timko et al. [26] was the failure of large-scale *Jatropha* investments to massively impact infrastructure growth and expansion. Except for the Jimle/Kpachaa investment destination, the *Jatropha* investment communities never benefited from the investors in terms of infrastructural provisions. This revelation runs parallel to the stance of Brittaine and Lutaladio [104] that *Jatropha* investment could trigger the provision of rural-based infrastructure for local transformation. In many of the *Jatropha* centers, jobs were created, but this did not have substantial economic improvements in affected communities. In fact, community members were generally worse off due to the large-scale *Jatropha* cultivation [105].

Jatropha investments also led to conflict of varying types [105] amongst the various interest groups, including investors, farmers, traditional authorities, and government agencies. The cause of the conflict was triggered by the following conditions: the lack of community participation and official notifications of the affected individuals during the periods of *Jatropha* investments; the meager and inconclusive compensation packages; the lack of transparency in the lease arrangements; the lack of trust in the overall negotiation process for *Jatropha* investments; ambiguous land ownership structure; and the lack of alternative productive agricultural lands for dispossessed farmers. Acheampong and Campion [105] vied that Ghana's *Jatropha* investment outcomes in the initial stages were indications that large-scale *Jatropha* plantations may not help the country attain the proposed ecological and livelihood benefits from *Jatropha* as widely spread in secondary sources.

4. Discussion

This review shows that, in all of the previously *Jatropha*-producing countries considered, none of them were committed to initiating scientific studies on the crop before jumping into its investment [47]. The general enticing attributes granted to the crop were enough to lure these countries to invest in the crop, thus placing huge expectations on the crop as a "miracle" to solve most of their development expectations. Indications including Openshaw's [2] provisions that *Jatropha* could rapidly spearhead economic growth provided a sound ground for the countries to jump into its cultivation. The motivations of countries were manifested through the initiation of biofuel policies with various levels of targets within a specified period. However, as the biofuel policies of the six reviewed countries hinged on specific investment opportunities in favor of green fuel in such countries, the central and common reason for the quick jump to *Jatropha* investment without commensurate and proper empirical research works was due to the quest to become energy autonomous.

The researchers were not totally pessimistic about *Jatropha*, as there were wide expectations for the young crop in a new system with no previous extensive practically scientific proven research. There was no local level research to prove country-specific conditions and the suitability of *Jatropha*, genetic composition, characteristics of *Jatropha*, its soil requirements, suitability, agronomy practices, marketability, and other undisclosed information about the crop. In Ethiopia, Mozambique, and Ghana, where land is the greatest asset for rural livelihoods, there was no critical consideration granted to land issues in terms of how best to incorporate *Jatropha* investments without local land denial and subsequent land alienation. The research, thus, acknowledged that the time for *Jatropha* investment was generally wrong, since it was the period for in-depth studies and knowledge-sharing amongst countries before gradual investments were started. The period was. thus, "research to prove before production" instead of the "wait-and-see approach after production" [106], which was an "easy way out and a short-cut to meet doubtful expectations". The wait-and-see approach, according to Gordon et al. [107], is a deferment approach. The deferment approach is due to the uncertainties associated with the potential allocation of information breaches. This uncertainty is the result of the potential vulnerabilities and threats associated with breaches. Due to these uncertainties, it may be rational to take a wait-and-see approach. In the case of Mexico, it was obvious that the local systems of *Jatropha* investment needed improvement through training and capacity building before the actual *Jatropha* investment, but this never took place. In the Indian case, the government was not financially prepared for *Jatropha* investments, and vague promises were made without fulfilment; as such, no one expected positive outcomes from the *Jatropha* investment in the country. The Chinese case also saw limited support offered by the government to corporate *Jatropha* investors, which led them to pull out from the investments. Perhaps, if the government had positioned itself to substantially support corporate bodies through subsidies during the investment, the development outcomes would have been positive. In Ethiopia, Mozambique, and Ghana, foreign investors rushed to invest without giving recognition to the traditional land rights of the locals. Without this social license from the tradition land-owners and users, instituting a successful business will always be contentious without local support. Figure 2 presents a summary of the global transformation creation trajectory through *Jatropha* investment and its diversity based on the approach to investment.

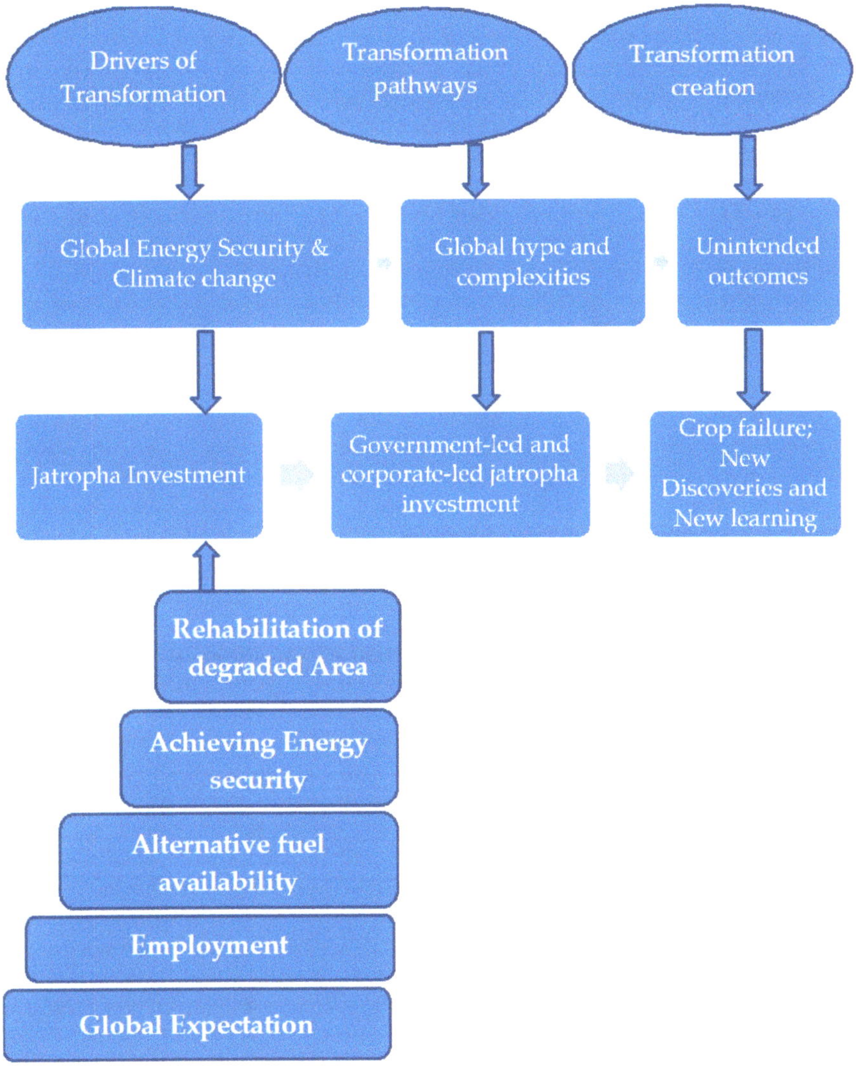

Figure 2. Global transformation trajectory through *Jatropha* investment.

Transformations in Government- versus Corporate-Led Jatropha *Investment*

Transformations of *Jatropha* investment took two main forms, that is, production initiated by the government and production initiated by corporate or private investors. In all six countries considered in the research, *Jatropha* production failed whether it was led by the government or by corporate investors. In Mexico and India, *Jatropha* investment was a national agenda with the government taking direct involvement in the cultivation process. The Chinese embraced a mix of socialist and capitalist approaches, but the government was somewhat reluctant to inject subsidies into the *Jatropha* initiative. Although Ethiopia and Mozambique have a centralized government control system, the corporate agenda (capitalist approach) was embraced. In Ghana, it was corporate-led in a decentralized democratic system. Based on the corporate production approach adopted by these countries, there were diversities in terms of the environmental, economic, and social transformations after the collapse of *Jatropha*. From the narrative, government involvement (socialist approach) tried to put "people" at the center of the investment, whilst corporate investors (capitalist approach) placed emphasis on "profit". In relation to the differences in priorities, government-led production focused on "marginal lands", instead of "productive lands" occupied by indigenous and tenant farmers [48,63,69,74,93]. Thus, the rate of environmental destruction for government-led production was lower than that of corporate investors. In fact, government-led production tried putting environmental integrity at the core of its investment. In India, for instance, Zafar [63] reiterated that the government involvement in *Jatropha* production was aimed at reducing soil erosion and the rehabilitation of land for the benefit of communities. The use of marginal lands in Mexico by the government also meant that *Jatropha* cultivation would not consume the green vegetation of communities [57]. In the Yucatan State, Mexico, the government tied *Jatropha* production to afforestation/reforestation with some hectares of land used specifically to expand forest. Chinese government engagement in the investment also led to extensive use of marginal lands, thus safeguarding productive land for the production of food crops.

In relation to corporate investors, the top priority of "profit" meant that lands as livelihood assets for host communities were taken away from them, and forested landscapes were transformed to large-scale *Jatropha* plantations, as in the case in Ethiopia, Mozambique, and Ghana. Flexibility in the use of abandoned lands by both the government and corporate investors varied sharply. In most of the government-led projects, local farmers were advised to use their lands for *Jatropha* production. After the poor outcomes of the initial investment, farmers later transformed their lands back to food crop production with ease. The marginal lands designated by the government for *Jatropha* lay idle, probably due to the fear of low fertility to support farming or the complexities associated with using government lands for personal projects [107]. In Mexico, India, and China, for instance, where there was some active form of government participation, lands used for the investment still lie dormant. With the corporate investors, once they were unable to realize gains from their investment, they left their lands dormant, and some community members ended up using such land for their own farming. In Ethiopia, Mozambique, and Ghana, where investments were predominantly led by private investors, some farmers returned to the *Jatropha* lands for farming purposes [81,86,88]. The subsequent re-transformation of land back to farmland helped advance the livelihood of such farmers, although they can be alienated at any point in time, especially when another investment sets in. There were also instances where corporate lands were re-leased to other private investors for investment; hence, farmers still faced alienation despite the bust of *Jatropha* projects in their communities.

The social implications associated with government- versus private-led investment varied. Whilst land-resource conflict became the hallmark of private investment due to alienation and unfair compensation, government-led investment intensified social capital, helping concretize social cohesion. In Ethiopia, Mozambique, and Ghana, where investment was intensified by private actors, issues on land rights became critical, leading to investor–farmer conflicts in the host communities [73,83,105]. This conflict affected social relations, and somewhat interrupted the operational peace investors needed for sound investment. In the case of government-led investments, government-owned lands were used; hence, issues of eviction were negated [55]. Again, farmers were encouraged to use their own lands for *Jatropha* cultivations. This solidified the unity of work, as farmers had to coordinate efforts and share responsibilities for the cultivation of a common crop, *Jatropha*. As a result, government-led investment did not create social disturbances vis-à-vis private investments.

In countries where *Jatropha* projects were led by the government, particularly in India, the government's intentions and plans were to encourage local farmers through incentive-driven approaches, even though they were not successful. However, the government continually provided the leading role, allowing farmers and agro-business groups to use their own resources for investment. *Jatropha* was planted on a large-, medium-, and small-scale basis, depending on the financial viability of the entities and individuals involved. Ethiopia, Mozambique, Ghana, and China, to a certain extent, saw the direct participation of corporate bodies in *Jatropha* investment. These corporate bodies were both local and foreign-based entities. The greatest proportion of foreign investors (corporate bodies) featured their investment in these countries due to the government's "open-door policy". Unlike in Mexico and India, *Jatropha* in Ethiopia, Mozambique, and Ghana was basically instigated on a large scale, with investors acquiring large tracts of land for their investments. Although there was land-use change, issues regarding land takeovers and the recognition of local land rights were rare in Mexico, India, and China due to the wide use of marginal lands (especially in Mexico and China) and the farmers' decision to use their own lands for *Jatropha* (India). In contrast, issues of land-use rights and tenure security became a critical development issue in Ethiopia, Mozambique, and China, as many local farmers were evicted from their lands without their approval or commensurate compensation. This affected local agriculture, with direct negative implications on food security. Concerns about food insecurity were also common in Mexico and India, where new pests and diseases were introduced, thus affecting food crops. Particularly in Mexico, the case was further worsened by some of the farmers deciding to vacate crop production for *Jatropha* cultivation. China very rarely had issues concerning the negative outcome of *Jatropha* on food production as marginal lands were predominantly used for *Jatropha* cultivation.

Jatropha investment led to the creation of temporal jobs for the local actors involved, but this did not really enhance the livelihood outcomes. In all six cases, *Jatropha* was unprofitable; as such, the target of meeting global energy needs through biofuel was dashed. This led to the *Jatropha* bust, with governments shifting their policy focus from *Jatropha* to make room for evidence-based research, refocusing their attention on a wide range of possible agro-fuel crops. Actors, including governments, local agro-business groups, farmers, and foreign investors, largely withdrew their investment due to the unprofitability of *Jatropha*. In Mexico, India, Ethiopia, Ghana, and Mozambique, some acres of *Jatropha* lands are used for the cultivation of other crops either by farmers or other investors, whilst other portions are yet to be used. Since marginal lands were used for *Jatropha* in China, the lands are still idle after the cessation of investment. Generally, *Jatropha* failed because of the quick leap to its investment without commensurate empirical studies, especially at country-specific levels to prove/disprove the heightened proclamation on the crop. *Jatropha* was just an ordinary crop, instead of the fancy names given to it such as "wonder crop", "miracle crop", etc.

The ongoing discussions imply that interventions based on lessons from both government-led and private-initiated investment need variations to critically address the outcomes associated with large-scale investment projects. There is a need for critical scrutiny and policy-based initiatives to comprehensively address the problems associated with large-scale investments either initiated by the government or by private individuals.

Currently, these countries have come to terms with the uncertain path of *Jatropha* transformation. Transformation patterns ultimately followed policy, investment, and land-use transformations (see Table 1). Thus, biofuel policy attention shifted from *Jatropha* to other feedstocks, investment toward *Jatropha* transformed to other crops or to no investment, and land use for *Jatropha* was transformed to farming grounds by either smallholders/other investors or to total abandonment without any crop production. *Jatropha* transformation led to socio-economic and environmental outcomes, with their diversities widely dependent on the perspective from which such investments were tackled, either led by the government (Mexico and India), corporate investors (Ethiopia, Mozambique, and Ghana), or concurrently by the government and corporate investors (China). As inferred from Deming [30], and Dazko and Sheinberg's [31] theory of transformation creation, the socio-economic and environmental outcomes of *Jatropha* transformation in these six countries as the initial uncertain destinations led to new discoveries and learning, especially toward an effective response to inform future investments. These countries have, therefore, come to terms with the outcomes of investment transformation (system) change through the path of transformation; thus, vast potential lessons exist for them to tap into so as to drive future agro-investment in a positive direction. Table 1 shows a summary of the reviewed literature on transformation creation through global *Jatropha* investment of the six selected previous *Jatropha*-producing countries, and their patterns of transformation vis-à-vis policy, investment, and land use.

Table 1. Global perspective of transformation creation in different *Jatropha* centers.

Country	Support Sources: Government- versus Corporate-Led *Jatropha* Investment	Transformation			
		Policy Transformation	**Environment, Social, and Economic Transformation**	**Investment Transformation**	**Land-Use Transformation**
Mexico	- Government-led - Investors/ agro-industries	National Biofuel Policy attention shifted from *Jatropha*.	- Land: Limited negative outcomes recorded on tenure issues. - Food security: Food production affected especially maize and beans; new pests and diseases. - Employment: Jobs created through agro-industries. - Energy security: Failure to meet biofuel targets.	- Withdrawal of investment by both government and investors.	- Farmers diverted the land to cultivation of food crops. - Marginal lands are idle.
India	- Government-led support (but this turned to be limited) - Farmers' own resources - Finance from local enterprises	National Biofuel Policy attention shifted from *Jatropha*.	- Land: Limited land tenure issues. - Food security: Pests and diseases introduced via *Jatropha* which affected food crop production. - Employment: Jobs created via *Jatropha* production at the local level. - Energy security: Failure to meet biofuel targets.	- Withdrawal of investment by both government and investors.	- Community members took over some lands whilst others are yet to be utilized.
China	- Corporate-led - Investors in collaboration with the government	National Biofuel Policy attention shifted from *Jatropha*.	- Land: Limited land tenure issues. - Food security: Food production and availability were not affected by *Jatropha*. - Employment: Jobs were created through contract farming. - Energy security: Failure to meet biofuel targets.	- Investment withdrawn by investors and government's interest dwindled.	- Lands lie dormant.
Ethiopia	- Corporate-led - Government support Foreign direct investment (FDI)	National Biofuel Policy attention shifted from *Jatropha* toward resilient Policy on Agriculture Modernization	- Land: Issues on land-use rights and tenure security became a critical development concern. - Food security: *Jatropha* somewhat affected food production as grazing areas and farmlands used for rotational farming were alienated. - Employment: Contract jobs were created, but payments were neither fair nor competitive. - Energy security: Failure to meet biofuel targets.	- Investment on halt in some areas. - Investment shifted to other crops in other areas.	- Lands used for other crops by investors. - Farmers took over their lands.

Table 1. *Cont.*

Country	Support Sources: Government- versus Corporate-Led *Jatropha* Investment	Transformation			
		Policy Transformation	Environment, Social, and Economic Transformation	Investment Transformation	Land-Use Transformation
Mozambique	- Corporate-led - Government support - Corporate funding	National Biofuel Policy attention shifted from *Jatropha*.	- Land: Issues on land-use rights and tenure security became a critical development concern. - Food security: Mixed results. Pests and diseases introduced via *Jatropha* which affected food crop production in some cultivated parts. - Employment: Jobs were created via corporate bodies. - Energy security: Failure to meet biofuel targets.	- Investment diverted to other crops by investors. - Other investors completely withdrew their investments.	- Lands used for other crops by investors. - Farmers took over their lands.
Ghana	- Corporate-led - Government support - Foreign direct investment (FDI) - Non-Governmental Organization (NGO) support	National Biofuel Policy attention shifted from *Jatropha* toward resilient Policy on Agriculture Modernization	- Land: Issues on land-use rights and tenure security became a critical development concern. - Food security: *Jatropha* somewhat affected food production as grazing areas and farmlands used for rotational farming were alienated. - Employment: Contract jobs were created, but payments were neither fair nor competitive. - Energy security: Failure to meet biofuel targets.	- Investment diverted to other crops by investors. - Other investors completely withdrew their investments.	- Lands used for other crops by *Jatropha* investors/new investors. - Farmers took over their lands but with no tenure security on land.

5. Conclusions

The global recognition given to *Jatropha* as a "wonder crop" was clearly a bust. The major investment destinations reviewed in this paper failed to achieve their expected results in terms of alleviating local and global energy poverty, reducing the greenhouse gas (GHG) emissions, and enhancing local livelihoods and development. The review established that the intended goal for establishing global *Jatropha* investment, which was to serve as an alternative source of fuel, failed because of the unexpected complexities of the uncertainties and the ubiquitous nature of the hype, which dwelled on the wait-and-see (that is, the deferment option) approach for global *Jatropha* investment. These frustrations caused governments and investors to lose interest in *Jatropha*. It should be accepted that *Jatropha* was over-hyped, in the absence of extensive and convincing scientific research works. *Jatropha* will not gain popular attention any longer, as countries seem reluctant to massively pay credence to its re-investment. Notwithstanding, the current phase of *Jatropha* transformation in the studied countries calls for them to move forward, as countries need to appreciate diversity in policy interventions in relation to key factors that lead to the investment implementation processes. This is premised on the fact that the outcomes of the *Jatropha* transformations in the countries studied are parallel to that which was premised, but not based on the approach and perspective to which *Jatropha* investments were tackled. Differences in socio-economic and environmental outcomes call for these countries to reflect upon the existing potentials and constraints created for responsive policy measures toward their biofuel sectors. Currently, these countries have transformed their biofuel policy attention from *Jatropha* to other feedstocks, and it is expected that these transformed policies will take precautions based on lessons learnt from the previous investments, as well as the current situations that are presented as an outcome through their respective travels on the path of transformation [30,31]. We support the shift in policy attention, since we are not all that pessimistic about the economic and production viability of *Jatropha* (even if treated with adequate responsibility). Perhaps considerable attention granted to other feedstocks, as these countries adopted, might have a future as a better alternative source of energy. It is expected that countries which embrace similar investment approaches, for instance, Mexico and India for government-led investment, and Ethiopia, Mozambique, and Ghana for corporate-led investment, might have some interrelated responses. Policy measures should, however, deeply reflect the peculiar dynamics in each country for effective outcomes.

Our research incites policy lessons that the adoption and promotion of any new crop must be based on incremental measures for sound policy responsiveness and implementation. The "quick jump" to any new crop by countries just because it is widely propagated by the international community as a response to a particular development need is not adequate enough for mere acceptance. Thought-provoking policy measures are needed, premised on wide consultation and a participatory paradigm characterized by in-depth empirical research within a particular country. As suggested by Muys et al. [36], policy measures toward the adoption and promotion of new crops should hinge on viable data from proposed investment communities, as well as a cost–benefit analysis of the outcomes of such investments. Even if this proves satisfactory, initial investment based on experimental trials in selected locations should be embraced as the first step, as this can help to unravel peculiar uncertainties and problems associated with such crops, as suggested by Soto et al. [108]. Another insightful policy implication of these findings is centered on policy interconnectedness, and the need for countries to be aware of and appreciate such connections for effective measures. Lempert [37] revealed that a single policy implementation is usually insufficient to tackle a particular development problem. Therefore, there is a need for national governments, corporate investors, and other important stakeholders to understand that the policy environment is a system made up of a network of several policies that work interrelatedly to address a particular development need [109]. This policy system is also applicable to the emergence and promotion of a new crop, which is likely to present solutions to a development problem. Ultimately, government and private investors must sufficiently reflect on the suitability of endorsing new investment initiatives before accurately appraising the investment viability. Even if there is such an endorsement, the aftermath of the failures of such initiatives should be appraised to understand the impacts of the failure to transform the system.

Author Contributions: The lead author, R.A.-B., conceived the idea and prepared the manuscript under the supervision of, and with contributions from A.Z., K.O., and A.A.

Funding: This project was funded by the Netherlands Organization for Scientific Research (NWO).

Acknowledgments: This project was supported by the Netherlands Organization for Scientific Research (NWO). The funding was granted as part of a program dubbed Conflict and Cooperation over Natural Resources (CoCooN).

Conflicts of Interest: The authors declare no conflicts of interest.

Notes: UN-Energy is a collaborative framework for all United Nations (UN) bodies that contributed to energy solutions. It was born out of the 2002 World Summit on Sustainable Development (WSSD) in Johannesburg, South Africa.

References

1. Sulle, E.; Nelson, F. *Biofuels, Land Access and Rural Livelihoods in Tanzania*; IIED: London, UK, 2009.
2. Openshaw, K. A review of Jatropha curcas: An oil plant of unfulfilled promise. *Biomass Bioenergy* **2000**, *19*, 1–15. [CrossRef]
3. Achten, W.M.; Maes, W.H.; Aerts, R.; Verchot, L.V.; Trabucco, A.; Mathijs, E.; Singh, V.P.; Muys, B. Jatropha: From global hype to local opportunity. *J. Arid Environ.* **2010**, *74*, 164–165. [CrossRef]
4. Achten, W.M.; Mathijs, E.; Verchot, L.; Singh, V.; Aerts, R.; Muys, B. Jatropha biodiesel fuelling sustainability? *Biofuels Bioprod. Biorefin.* **2007**, *1*, 283–291. [CrossRef]
5. Achten, W.M.J.; Verchot, L.; Franken, Y.J.; Mathijs, E.; Singh, V.P.; Aerts, R.; Muys, B. *Jatropha* bio-diesel production and use. *Biomass Bioenergy* **2008**, *32*, 1063–1084. [CrossRef]
6. Bassey, N. The Agrofuels Debate in Africa: Challenges and Opportunities. In Proceedings of the Ecological Agriculture: Mitigating Climate Change, Providing Food Security and Self-Reliance for Rural Livelihoods in Africa Conference, Addis Ababa, Ethiopia, 26–28 November 2008.
7. Jingura, R.M.; Matengaifa, R.; Musademba, D.; Musiyiwa, K. Characterization of land types and agro-ecological conditions for production of Jatropha as a feedstock for biofuels in Zimbabwe. *Biomass Bioenergy* **2011**, *35*, 2080–2086. [CrossRef]
8. Tsegaye, W.; Glantz, M.H. *Biofuels in Africa: A Path Way to Development*; Occasional Paper No. 43; International Research Center for Energy and Economic Development: Boulder, CO, USA, 2011.
9. Timko, J.A. An Analytical Framework for Assessing the Impacts of Jatropha Curcas on Local Livelihoods. In *Conflict and Cooperation over Natural Resources in the Global South: Conceptual Approaches*; Bavinck, M., Pellegrini, L., Mostert, E., Eds.; CRC Press, Taylor and Francis Group: Boca Raton, FL, USA, 2014; pp. 173–191.
10. Bosch, C.; Zeller, M. The impacts of wage employment of jatropha plantation on income and food security of rural households in Madagascar—A panel data analysis. *Q. J. Int. Agric.* **2013**, *52*, 119–140.
11. Food and Agriculture Organization. *The Gender and Equity Implications of Land-Related Investments on Land Access, Labour and Income-Generating Opportunities in Northern Ghana: The Case Study of Integrated Tamale Fruit Company*; FAO: Rome, Italy, 2013.
12. Dias, L.A.S.; Missio, R.F.; Dias, D.C.F.S. Antiquity, Botany, Origin and Domestication of Jatropha Curcas (Euphorbiaceae), a Plant species with potential for biodiesel production. *Genet. Mol. Res.* **2012**, *11*, 2719–2728. [CrossRef]
13. Leonti, M.; Sticher, O.; Heinrich, M. Antiquity of medicinal plant usage in two Macro-Mayan ethnic groups (Mexico). *J. Ethnopharmacol.* **2003**, *88*, 119–124. [CrossRef]
14. Henning, R.K. *The Jatropha Booklet: A Guide to Jatropha Promotion in Africa*; Bagani GbR: Weissensberg, Germany, 2003; pp. 5–33.
15. Nogueira, L.A.H. *Perspectivas de un Programa de Biocombustibles en América Central*; CEPAL/GTZ: Mexico City, Mexico, 2004. Available online: https://repositorio.cepal.org/bitstream/handle/11362/25669/LCmexL606_es.pdf?sequence=1&isAllowed=y (accessed on 1 June 2019).
16. Bassey, N. Agrofuels: The corporate plunder of Africa. *Third World Resur.* **2009**, *223*, 21–26.
17. Hill, J.; Nelson, E.; Tilman, D.; Polasky, S.; Tiffany, D. Environmental, economic and energetic costs and benefits of biodiesel and ethanol biofuels. *Proc. Natl. Acad. Sci. USA* **2006**, *103*, 11206–11210. [CrossRef]

18. Graham, V.M.; Gasparatos, A.; Fabricius, C. The Rise, Fall and Potential Resilience Benefits of Jatropha in Southern Africa. *Sustainability* **2014**, *6*, 3615–3643. [CrossRef]
19. Food and Agriculture Organization (FAO). *The State of Food and Agriculture. Investing in Agriculture for a Better Future*; International Institute of Environmental Development (IIED): London, UK, 2012.
20. Hughes, A.K.; Knox, A.; Jones-Casey, K. *Focus on Land Brief in Africa: Pressure on Land from Large Scale Biofuel Production. Lesson 2*; Funded by the Bill and Melinda Gates foundation. IFAD Occasional Paper 2; World Resources Institute in Partnership with Landesa Rural Development Institute: Washington, DC, USA, 2011.
21. Schoneveld, G.C.; German, L.A.; Nutakor, E. *Towards Sustainable Biofuels Development: Assessing the Local Impacts of Large-Scale Foreign Land Acquisitions in Ghana*; World Bank: Washington, DC, USA, 2010; pp. 1–7.
22. Action Aid. *Rethinking the Rush to Agrofuels: Lessons from Ghana, Senegal and Mozambique on the Unintended Consequences of Agrofuels Production for Food Security*; Action Aid: Johannesburg, South Africa, 2009.
23. Cotula, L.; Dyer, N.; Vermeulen, S. *Fuelling Exclusion? The Biofuels Boom and Poor People's Access to Land*; IIED: London, UK, 2008.
24. Skutsch, M.; de los Rios, E.; Solis, S.; Riegelhaupt, E.; Hinojosa, D.; Gerfert, S.; Gao, Y.; Masera, O. Jatropha in Mexico: Environmental and social impacts of an incipient biofuel program. *Ecol. Soc.* **2011**, *16*, 11. [CrossRef]
25. Birega, G. Agrofuels Beyond the Hype: Lessons and Experiences from other countries. In *Agrofuel Development in Ethiopia: Rhetoric, Reality and Recommendations*; Heckett, T., Aklilu, N., Eds.; Forum for Environment: Addis Ababa, Ethiopia, 2008; pp. 67–83.
26. Timko, J.A.; Amsalu, A.; Acheampong, E.; Teferi, M.K. Local Perceptions about the Effects of Jatropha (Jatropha curcas) and Castor (Ricinus communis) Plantations on Households in Ghana and Ethiopia. *Sustainability* **2014**, *6*, 7224–7241. [CrossRef]
27. Hamenoo, S.V.Q. The Effects of Large-Scale Land Acquisition for Jatropha Plantation on Small-Scale Farmers in Rural Communities in the Asante Akim North District. Master's Thesis, School of Graduate Studies Kwame Nkrumah University of Science and Technology, Kumasi, Ghana, 2014. Available online: http://dspace.knust.edu.gh/bitstream/123456789/7157/1/HAMENOO%2C%20SIMON%20VICTORY%20QUARCSON.pdf (accessed on 10 July 2018).
28. Aha, B.; Ayitey, J.Z. Biofuels and the hazards of land grabbing: Tenure (in) security and indigenous farmers' investment decisions in Ghana. *Land Use Policy* **2017**, *60*, 48–59. [CrossRef]
29. Farazmand, A. Chaos and transformation theories: A theoretical analysis with implications for organization theory and public management. *Public Organ. Rev.* **2003**, *3*, 339–372. [CrossRef]
30. Deming, W.E. *The New Economics*; MIT Press: Cambridge, UK, 1993.
31. Daszko, M.; Sheinberg, S. Survival is optional: Only leaders with new knowledge can lead the transformation. *Transformation* **2005**, *408*, 247–7757.
32. Murphy, J.; Hallinger, P. The principalship in an era of transformation. *J. Educ. Adm.* **1992**, *30*. [CrossRef]
33. Sarasvathy, S.D.; Dew, N. New market creation through transformation. *J. Evol, Econ.* **2005**, *15*, 533–565. [CrossRef]
34. Gleick, J. *Chaos: Making a New Science*; Knopf: New York, NY, USA, 1987.
35. Wheatley, M. *Leadership and the New Science: Discovering Order in a Chaotic World*, 2nd ed.; Berrett-Koehler: Washington, WA, USA, 1999.
36. Muys, B.; Norgrove, L.; Alamirew, T.; Birech, R.; Chirinian, E.; Delelegn, Y.; Ehrensperger, A.; Ellison, C.A.; Feto, A.; Freyer, B.; et al. Integrating mitigation and adaptation into development: The case of Jatropha curcas in sub-Saharan Africa. *Glob. Chang. Biol. Bioenergy* **2014**, *6*, 169–171. [CrossRef]
37. Lempert, R.J. *Shaping the Next One Hundred Years: New Methods for Quantitative, Long-Term Policy Analysis*; Rand Corporation: Santa Monica, CA, USA, 2003.
38. Brenner, R. The agrarian roots of European capitalism. *Past Present* **1982**, *97*, 16–113. [CrossRef]
39. Kenney-Lazar, M. Plantation rubber, land grabbing and social-property transformation in southern Laos. *J. Peasant Stud.* **2012**, *39*, 1017–1037. [CrossRef]
40. Keijzer, N.; Lundsgaarde, E. *When Unintended Effects Become Intended: Implications of 'Mutual Benefit' Discourses for Development Studies and Evaluation Practices; Working Paper*; Ministry of Foreign Affairs of the Netherlands: Hague, The Netherlands; Radboud University: Nijmegen, The Netherlands, 2017.
41. Koch, D.J.; Schulpen, L. Unintended effects of international cooperation: A preliminary literature review. In Proceedings of the Unintended Effects of International Cooperation, Hague, The Netherlands, 16–17 January 2017.
42. Downe-Wamboldt, B. Content analysis: Method, applications, and issues. *Health Care Women Int.* **1992**, *13*, 313–321. [CrossRef]

43. Kondracki, N.L.; Wellman, N.S. Content analysis: Review of methods and their applications in nutrition education. *J. Nutr. Educ. Behav.* **2002**, *34*, 224–230. [CrossRef]
44. Morse, J.M.; Field, P.A. *Qualitative Research Methods for Health Professionals*, 2nd ed.; Sage: Thousand Oaks, CA, USA, 1995.
45. Miles, M.B.; Huberman, A.M. *Qualitative Data Analysis: An Expanded Sourcebook*; Sage: Thousand Oaks, CA, USA, 1994.
46. Patton, M.Q. *Qualitative Research and Evaluation Methods*; Sage: Thousand Oaks, CA, USA, 2002.
47. Banerjee, A.; Halvorsen, K.E.; Eastmond-Spencer, A.; Sweitz, S.R. Sustainable development for whom and how? Exploring the gaps between popular discourses and ground reality using the Mexican Jatropha biodiesel case. *Environ. Manag.* **2017**, *59*, 912–923. [CrossRef]
48. Montero, G.; Stoytcheva, M.; Coronado, M.; García, C.; Cerezo, J.; Toscano, L.; León, J.A. An overview of biodiesel production in Mexico. In *Biofuels-Status and Perspective*; InTech: London, UK, 2015.
49. Solomon, B.D.; Bailis, R. (Eds.) *Sustainable Development of Biofuels in Latin America and the Caribbean*; Springer Science & Business Media, 2013; Available online: https://link.springer.com/book/10.1007%2F978-1-4614-9275-7 (accessed on 10 June 2019).
50. IICA. *México—Inicia Yucatán Cuzltivo de Jatropha para Biodiesel*; IICA: New Delhi, India, 2010.
51. Zamarripa-Colmenero, A.; Diaz Padilla, G. *Areas de Potencial Productivo del Pinon Jatropha Curcus, L., Como Especie de Interés Bioenergético en Mexico*; Boletin No. 16; Oleaginosa: Mexico City, Mexico, 2008; pp. 4–6.
52. Robinson, S.; Beckerlegge, J. Jatropha in Africa: Economic Potential. 2008. Available online: http://www.wolfsberg.com/documents/Jatropha_in_Africa_Economic_Potential.pdf (accessed on 18 January 2019).
53. Hinojosa, F.I.D.; Skutsch, M. Impactof establishing jatropha curcas t produce biodiesel in three communities of Michocan, mexico, approached from different scales. *Rev. Geogr. Am. Cent.* **2011**, *2*, 1–15.
54. Rucoba, G.A.; Munguía, G.A.; Sarmiento, F.F. Between Jatropha and poverty: Reflections about biofuels production in temporary lands in Yucatan(Entre la Jatropha ylapobreza: Reflexiones sobre la producción de agrocombustibles en tierras de temporal en Yucatán). *Estud. Soc.* **2012**, *21*, 115–142.
55. Valero, P.J.; Cortina, V.S.; Vela, V.S. The project of biofuels in Chiapas: Experiences of physic nut (Jatropha curcas) farmers within the rural crisis framework. *Estud. Soc.* **2011**, *19*, 120–144.
56. Ariza-Montobbio, P.; Lele, S. Jatropha plantations for biodiesel in Tamil Nadu, India: Viability, livelihood trade-offs, and latent conflict. *Ecol. Econ.* **2010**, *70*, 189–195. [CrossRef]
57. Schut, M.; Slingerland, M.; Locke, A. Biofuel developments in Mozambique. Update and analysis of policy, potential and reality. *Energy Policy* **2010**, *38*, 5151–5165. [CrossRef]
58. Rodriguez, O.A.V.; Vazquez, A.P.; Gamboa, C.M. Drivers and Consequences of the First Jatropha Curcas Plantations in Mexico. *Sustainability* **2014**, *6*, 3732–3746. [CrossRef]
59. Sweitz, S. Sustainability, Biofuels, and the Future in Yucatán. In *The Quest for Jatropha Biodiesel and Sustainability in Yucatan*; Ediciones de la Universidad Autónoma de Yucatán: Mexico City, Mexico, 2018; p. 239.
60. Chan, C.J. En proceso el cultivo de 62,000 hectáreas de jatropha en Yucatán. *Diario* **2010**. (In Spanish). Available online: http://biodiesel.com.ar/3289/cultivo-de-jatropha-para-producir-biodiesel-en-mexico (accessed on 28 August 2018).
61. Global Agricultural Information Network (GAIN). *Biofuel Annuals: Uncertainty of the Future of Mexican Biofuels. Gain Report Number MX2507*; GAIN: Washington, DC, USA, 2012.
62. Axlesson, L.; Franzen, M. Performance of Jatropha Biodiesel Production and Its Environmental and Socio-Economic Impacts—A Case of Southern India. Master's Thesis, Department of Energy and Environment, Chalmers University of Technology, Göteborg, Sweden, 2010.
63. Zafar, S. Biodiesel Scenario in India. 2011. Available online: https://www.academia.edu/30151096/Biodiesel_Scenario_in_India (accessed on 22 October 2018).
64. Romijn, H.A. Land clearing and Green House Gas emissions from jatropha biofuels on African Miambo Woodland. *Energy Policy* **2011**, *39*, 5751–5762. [CrossRef]
65. Ministry of New and Renewable Energy, Government of India. National Policy on Biofuels. 2008. Available online: https://mnre.gov.in/file-manager/UserFiles/biofuel_policy.pdf (accessed on 5 January 2019).
66. Reinhardt, G.; Gartner, S.; Rettenmaier, N.; Munch, J.; Von Falkenstein, E. *Screening Life Cycle Assessment of Jatropha Biodiesel*; IFEU-Institute for Energy and Environmental Research Heidelberg Gmbh: Heidelberg, Germany, 2007.

67. Prueksakorn, K.; Gheewala, S.H. Energy and Greenhouse Gas Implications of Biodiesel Production from Jatropha curcas. In Proceedings of the 2nd Joint International Conference on Sustainable Energy and Environment (SEE 2006), Bangkok, Thailand, 21–23 November 2006.
68. Ministry of New and Renewable Energy. Remap Renewable Energy Prospects for India. 2017. Available online: https://www.irena.org/-/media/Files/IRENA/Agency/Publication/2017/May/IRENA_REmap_India_paper_2017.pdf (accessed on 3 February 2019).
69. Li, J.; Bluemling, B.; Mol, A.P.J.; Herzfeld, T. Stagnating Jatropha Biofuel Development in Southwest China: An Institutional Approach. *Sustainability* **2014**, *6*, 3192–3212. [CrossRef]
70. Weyerhaeuser, H.; Tennigkeit, T.; Yufang, S.; Kahrl, F. *Biofuels in China: An Analysis of the Opportunities and Challenges of Jatropha Curcas in South West China, ICRAF Working Paper Number 53*; ICRAF: Beijing, China, 2007.
71. Benge, M. *Assessment of the Potential of Jatropha Curcas, (Biodiesel Tree,) for Energy Production and Other Uses in Developing Countries*; Agroforestry Office, USAID: Laguna, Philippians, 2006. Available online: http://www.ascension-publishing.com/BIZ/jatropha.pdf (accessed on 1 January 2019).
72. Rulli, M.C.; Savioria, A.; D'Odorico, P. Globalland and water grabbing. *Proc. Natl. Acad. Sci. USA* **2013**, *110*, 892–898. [CrossRef]
73. Portner, B.; Ehrensperger, A.; Nezir, Z.; Breu, T.; Hurni, H. Biofuels for a Greener Economy? Insights from Jatropha Production in Northeastern Ethiopia. *Sustainability* **2013**, *6*, 6188–6202. [CrossRef]
74. Federal Democratic Republic of Ethiopia (FDRE). *Ethiopia's Climate-Resilient Green Economy. Green Economy Strategy*; Environmental Protection Authority: Addis Ababa, Ethiopia, 2011; Available online: https://theredddesk.org/resources/ethiopias-climate-resilient-green-economy-green-economy-strategy (accessed on 10 June 2019).
75. Ministry of Mines and Energy (MoME). *Biofuel Development and Utilization Strategy*; The Federal Democratic Republic of Ethiopia: Addis Ababa, Ethiopia, 2007.
76. Ethiopian Biofuels Development Directorate. *Assessment of Biofuels Projects Status in Ethiopia*; Melca Mahiber: Addis Ababa, Ethiopia, 2011. Available online: http://melcaethiopia.org/wp-content/uploads/2011/05/Eth_Biofuel_Assessment-Final.pdf (accessed on 2 March 2019).
77. Bossio, D.; Erkossa, T.; Dile, Y.; McCartney, M.; Killiches, F.; Hoff, H. Water Implications of Foreign Direct Investment in Ethiopia's Agricultural Sector. *J. Water Altern.* **2012**, *5*, 223–242.
78. International Land Coalition. *Global Assembly 2011*; ILC: Tirana, Albania, 2011.
79. Rahmato, D. *Land to Investors: Large-Scale Land Transfer in Ethiopia*; FSS Policy Debate Series No.1; Forum of Social Studies (FSS): Addis Ababa, Ethiopia, 2011.
80. Wendimu, M.A. Jatropha potential on marginal land in Ethiopia: Reality or myth? *Energy Sustain. Dev.* **2016**, *30*, 14–20. [CrossRef]
81. Von Maltitz, G.P.; Gasparatos, A.; Fabricius, C.; Morris, A.; Willis, K.J. Jatropha cultivation in Malawi and Mozambique: Impact on ecosystem services, local human well-being and poverty alleviation. *Ecol. Soc.* **2016**, *21*, 3. [CrossRef]
82. UN Energy. *Sustainable Bioenergy: A Framework for Decision Makers*; Henceforth UN-Energy: New York, NY, USA, 2007. Available online: http://www.fao.org/3/a1094e/a1094e00.pdf (accessed on 4 February 2019).
83. Mataveia, M. Biofuel Policy and Strategy for Mozambique. Presented at a Conference on 'Bioenergy for Sustainable Development in Africa—Lessons learnt from COMPLETE'; Available online: http://www.globalbioenergy.org/fileadmin/user_upload/gbep/docs/2009_events/CSD_side_event_NY/Mataveia_-_GBEP_CSD_side_event_140509.pdf (accessed on 10 June 2019).
84. Slingerland, M.; Schat, M. Jatropha Developments in Mozambique: Analysis of structural conditions influencing Niche-Regime Interactions. *Sustainability* **2014**, *6*, 7541–7563. [CrossRef]
85. Bos, H.L.; Slingerland, M.A.; Elbersen, W.; Rabbings, R. Beyond Agrification; twenty years of policy and innovation for non-food application of renewable resources in the Netherlands. *Biofuels Bioprod. Biorefin.* **2008**, *2*, 343–357. [CrossRef]
86. Cuvilas, C.A.; Jirjisa, R.; Lucas, C. Energy situation in Mozambique: A review. *Renew. Sustain. Energy Rev.* **2010**, *14*, 2139–2146. [CrossRef]
87. Batidzirai, B.; Faaij, A.P.C.; Smeets, E. Biomass and bioenergy supply from Mozambique. *Energy Sustain. Dev.* **2006**, *10*, 54–81. [CrossRef]

88. Namburete, S. Mozambique biofuels. In Proceedings of the African Green Revolution Conference, Oslo, Norway, 2006; Volume 31. Available online: http://mediabase.edbasa.com/kunder/yaraimages/agripres/agripres/j2006/m09/t01/0000443_2.pdf (accessed on 3 March 2019).
89. Van Eijck, J.; Rom, C.J.; Romijn, H.; Heijnen, S.; De Ruijter, F.; Jongschaap, R. *Jatropha Sustainability Assessment, Data from Tanzania, Maliand Mozambique*; NL Agency: Utrecht, The Netherlands, 2013.
90. Romijn, H.; Heijnen, S.; Colthoff, J.R.; De Jong, B.; Van Eijck, J. Economic and Social Sustainability Performance of Jatropha Projects: Results from Field Surveys in Mozambique, Tanzania and Mali. *Sustainability* **2014**, *6*, 6203–6235. [CrossRef]
91. Boamah, F. How and why chiefs formalize land use in recent times: The politics of land dispossession through biofuel investments in Ghana. *Rev. Afr. Polit. Econ.* **2014**, *41*, 406–423. [CrossRef]
92. Brew Hammond, A. Bioenergy for accelerated agro-industrial development in Ghana. In Proceedings of the Bioenergy Markets, West Africa Conference, Accra, Ghana, 27 October 2009.
93. Ahmed, A.; Kanton, S.; Godwin, K.; Rahim, A.A.; Salia, R.A. Biofuel development and large scale land acquisition in Ghana, implications for land use planning. *Int. J. Dev. Res.* **2014**, *4*, 2563–2571.
94. Amoah, O. *Jatropha: A Catalyst for Economic Growth in Africa*; An Official UNCTAD Document; Anuanom Industrial Bio Products Limited: Accra, Ghana, 2006. Available online: https://unctad.org/Sections/wcmu/docs/ditc_comb_Jatropha001_en.pdf (accessed on 25 December 2018).
95. Agyekumhene, J.K. Supporting the development of Jatropha farms with Microfinance for job and wealth creation, reduction in urban drift. In Proceedings of the UNCTAD Biofuels Workshop, Financing Biofuels with Special Emphasis on Jatropha and CDM, Accra, Ghana, 13–14 November 2006.
96. Energy Commission. *Draft Bioenergy Policy of Ghana*; Energy Commission: Accra, Ghana, 2010.
97. Technoserve. *Feasibility Study of Biofuel Production in Ghana: Assessing Competitiveness and Structure of the Industry's Value Chain*; Final Report; Technoserve: Washington, WA, USA, 2007. Available online: http://s3.amazonaws.com/zanran_storage/elliott.gwu.edu/ContentPages/2454285956.pdf (accessed on 18 August 2018).
98. Energy Commission. *Strategic National Energy Plan 2006–2020*; Energy Commission: Accra, Ghana, 2006.
99. Iddrisu, I.; Bhattacharyya, S.C. Ghana's bioenergy policy: Is 20% biofuel integration achievable by 2030? *Renew. Sustain. Energy Rev.* **2015**, *43*, 32–39. [CrossRef]
100. Boni, S. Indigenous Blood and Foreign Labor: The Ancestralization of Land Rights inSefwi (Ghana). In *Land and the Politics of Belonging in West Africa*; Kuba, R., Lentz, C., Eds.; E.J. Brill: Leiden, The Netherlands, 2006; pp. 161–186.
101. UNDESA. *Small-Scale Production and Use of Liquid Biofuels in Sub-Saharan Africa: Perspective for Sustainable Development*; UNDESA: New York, NY, USA, 2007.
102. Karlson, G.; Banda, K. *Biofuels for Sustainable Rural Development and Empowerment of Women: Case Studies for Africa and Asia*; Energia: Leusden, The Netherlands, 2009.
103. James, L. Theory and Identification of Marginal Land and Factors Determining Land Use Change. Master's Thesis, Department of Agricultural, Food, and Resource Economics, Michigan State University, East Lansing, MI, USA, 2010.
104. Brittaine, R.; Lutaladio, N. *Jatropha: A Smallholder Bioenergy Crop: The Potential for Pro-Poor Development*; Food and Agriculture Organization of the United Nations (FAO): Reading, UK, 2010; Volume 8.
105. Acheampong, E.; Campion, B.B. The Effects of Biofuel Feedstock Production on Farmers' Livelihoods in Ghana: The Case of Jatropha curcas. *Sustainability* **2014**, *6*, 4587–4607. [CrossRef]
106. Gordon, L.A.; Loeb, M.P.; Lucyshyn, W. Information security expenditures and real options: A wait-and-see approach. *Comput. Secur. J.* **2003**, *19*, 1–7.
107. Lambin, E.F.; Turner, B.L.; Geist, H.J.; Agbola, S.B.; Angelsen, A.; Bruce, J.; George, P. The causes of land use and land cover change: Moving beyond the myths. *Glob. Environ. Chang.* **2001**, *11*, 261–269. [CrossRef]
108. Soto, I.; Ellison, C.; Kenis, M.; Diaz, B.; Muys, B.; Mathijs, E. Why do farmers abandon jatropha cultivation? The case of Chiapas, Mexico. *Energy Sustain. Dev.* **2018**, *42*, 77–86. [CrossRef]
109. Kenis, P.; Schneider, V. Policy networks and policy analysis: Scrutinizing a new analytical toolbox. In *Policy Networks: Empirical Evidence and Theoretical Considerations*; Campus Verlag: Frankfurt, Germany, 1991; pp. 25–59.

sustainability

Review

A PESTLE Analysis of Biofuels Energy Industry in Europe

Spyridon Achinas [1,*], Johan Horjus [1], Vasileios Achinas [2] and Gerrit Jan Willem Euverink [1]

[1] Faculty of Science and Engineering, University of Groningen, 9747 AG Groningen, The Netherland; j.s.horjus.1@student.rug.nl (J.H.); g.j.w.euverink@rug.nl (G.J.W.E.)
[2] Institute for Life Science and Technology, Hanze University of Applied Sciences, Zernikeplein 11, 9747 AS Groningen, The Netherlands; v.achinas@st.hanze.nl
* Correspondence: s.achinas@rug.nl

Received: 26 September 2019; Accepted: 24 October 2019; Published: 28 October 2019

Abstract: Biofuels production is expected to be an intrinsic confluence to the renewable energy sector in the coming years under the European regulations for renewable energy. Key standpoints of the biofuels promotions are the reduction of national carbon emissions and rural deployment. Despite jubilant outlook of biofuels for sustainable development, research efforts still tend to link the biofuel industry and regional growth. The aim of this study is to explore and review the biofuels industry through a socio-political, techno-economic, legal and environmental (PESTLE) analysis approach, and discuss the interrelation between technological facets and sustainable deployment.

Keywords: renewable fuel standard; biofuel industry; rural development; sustainable deployment

1. Introduction

The drive to switch to an alternative transportation fuel is to reduce the dependency on oil and decrease the pressure on the environment [1–8]. The targets set in the RED by the European Commission are to support the development of biofuels [9–14]. Most of the legislations that are developed to achieve these targets are based on life-cycle assessments which often only account for greenhouse gas emissions (GHG) [15–17]. Several other limitations regarding the sustainability of the production, including the competition of feedstock cultivation with the food industry, create difficulties to achieve the short-term production goals of biofuels set by the European Commission [18–23]. This results in the need to develop standard production practices as well as licensing and certification of biofuel manufacture, so that all relevant environmental and social impacts can be regarded [24–27]. To create such production standards, the governing organizations, which are the European Commission and the governments in Europe, need to understand the production processes and its impact on sustainability. Even when policies directed towards a more sustainable production are in place, the production companies need to understand how to adhere to the legislation. They will need to increase their research and development into the energy production processes [28–31].

The biofuel industry is still developing, as the consumption of all biofuels in the European Union (EU) increased with 8% from 2016 to 2017 [32]. In the EU, the renewable energy directive (RED) is created as a policy to support the production and development of these renewable energies [13]. Especially for the transportation sector, the objective for all the EU countries is to have 10% renewable energies in 2020 [33]. Biofuels are considered as the key product to achieve this target and further criteria regarding their sustainability are developed [34]. Several literature sources stress the importance of understanding the role of biofuels in the possible reductions in life-cycle greenhouse gas (GHG) emissions that can be achieved with respect to fossil fuels to inform policy development and the decisions regarding the best fuel types [35]. This research will take this broader by analyzing the sustainability factors which influence the biofuel industry, and relates this to the biofuel characteristics, including the type of fuel, feedstock, and conversion process. In addition to the support for policy development, the research and discussion of the biofuel characteristics have the intent to provide a foundation for the directives that companies are going to take to adhere to the policies. Sustainability assessment of biofuels production from biomass and biowaste is an important prerequisite for informed and sound decision-making. However, life-cycle analysis (LCA) is still constrained by the difficulty of pointing out the most relevant impact factors. To underwrite the mighty risks confronted by the industrial efforts are pointed out in order to bestow the optimal use of energy and resources. Identification of key stakeholders and deployment of the political, economic, social, technological, legal, and environmental aspects from the updated literature, reports, and guidelines are intrinsic to achieve a thorough understanding of the complex landscape of biofuel industry. The contribution of the biofuel industry to sustainable development is key in the growing biofuel industry that is driven by the renewable energy targets of the European Union. Both policymakers and production companies require an understanding of the relation between biofuel production characteristics and the sustainability of the industry to be able to support sustainable development [36–44]. The objective of this report is to (1) summarise the key interdependencies in the biofuel industry based on PESTLE analysis and (2) create a review in which the key biofuel production characteristics are correlated with the sustainability factors that are affected by the biofuel industry. Hereafter, the different aspects influencing the sustainability of bioethanol, biodiesel, and biomethane industry are determined and these are allocated to the economic, environmental, socioecological, and geopolitical facets [45–55]. Finally, both quantitative and qualitative data are assessed to correlate both aspects and determine the strengths and weaknesses of the different biofuel characteristics. The analysis of the production processes shows that the key factors to be considered are both the choice of feedstock and the conversion technology that is applied. A key distinction in the feedstock is the cultivation of crops dedicated to biofuel production or using wastes and residues from other sectors as feedstock. Regarding the conversion processes, all three biofuels can be produced via a biochemical or the thermochemical pathway [56–58]. Both feedstock and production process influence the economic sustainability, showing that crop-based biofuels have a high feedstock cost and low capital cost, compared to the relatively low cost for feedstock and high capital cost for the biofuels from lignocellulosic biomass, wastes, and residues [59–76]. Moreover, the latter are also significantly more environmental and socioecological sustainable compared to the crop feedstock that imposes the impacts of land use and land-use change. Geopolitically, the EU supports the EU biofuel producers by protecting the market from lower prices introduced by imported biofuels. The political support for the more extensive development that is required for the biofuels from lignocellulosic biomass, wastes, and residues is starting to increase, but significant differences are detected among EU countries due to a lack of budget and existing infrastructure. Finally, a small discussion about the future of the biofuel industry is provided, considering the future use of fossil fuel and the development of other renewable alternatives such as hydrogen fuel and electric vehicles [77–79].

It is important to stress that this research will not be directed towards one answer, but rather assesses collected knowledge to provide an overview of the interrelations between the production industry and the sustainability and highlight the apparent distinctions and characteristics. This research correlates the different biofuel production characteristics with the sustainability of the biofuel industry in the context of economic, environmental, socioecological, and geopolitical facets. Hereafter, both subjects will be combined with a thorough assessment and discussion of literature and data obtained from literature to create a correlation. Finally, an outlook will be provided to evaluate possible external influences on the biofuel industry in the future.

2. Materials and Methods

2.1. Problem Context

The problem context indicates a research interest in the sustainability of the biofuel production processes. Understanding the interrelation between the two is essential for the future of biofuels since it is its sustainability, both economically and environmentally, which has created greater attention for biofuels among the many other alternative energy sources [80–90]. Since the possibilities in production processes depend on the type of feedstock used in the processes, several feedstocks are taken into account [91–94]. To further limit the number of different production processes, only the three biofuels will be considered: Bioethanol, biodiesel, and biomethane.

The scope of the sustainability aspects is discussed by specifying the key issues of the sustainable development of biofuels which are divided into four subsections: Economic, environmental, socioecological, and geopolitical. The first three sections combined provide the sustainability factors involved with biofuel production as found in literature and the subdivision is supported by the framework of Mangoyana et al. (2013) [95]. Since the social factor introduced in the framework includes aspects such as land use and biodiversity, the term 'socioecological' is deemed more suited. Finally, the geopolitical substratum is added to consider the effect of governmental support and their policies on biofuel development. The support still differs between countries in the EU due to their own individual interests [14]. Moreover, the independence of energy supply is a driving force for adopting biofuels in the EU.

The factors influencing the economic sustainability of the biofuel industry are all related to the economic feasibility of production [96]. The key factor is the start of new biofuel production companies, or the extension of current production lines, for the development of the biofuel industry in terms of its market size with respect to other fuels. Hence, the focus will be on the choices in production processes, including the type of biofuel and feedstock, and its relation to the sustainability facets. The environmental facet includes the GHG emission of the process to the discussion of sustainability. The emissions of fossil fuels are an important driving force for the development of biofuels, which are expected to have a lower net GHG emission due to the plant feedstock [97–101]. In addition, there are the factors that have both an impact on the environment, the society, and the quality of life of the individual inside that society, that are covered in the socioecological facet [102].

The inputs of the system are the commercially applied production processes of bioethanol, biodiesel, and biomethane together with the factors that determine the economic, environmental, socioecological, and geopolitical sustainability of the production processes that will be determined and will be discussed in the methodology of this research. It is projected to deliver insights into the performance of the different production technologies, feedstocks, and final biofuel product with respect to the sustainability of the biofuel industry. First, the different possibilities in the production of bioethanol, biodiesel, and biomethane have to be determined. Thereafter, it has to be determined how different biofuel production characteristics contribute to the sustainability facets described in the scope of the research and summarized in the economic, environmental, socioecological, and geopolitical facets.

The main sources of information will be review articles on biofuel technologies to provide an overview of the existing technologies. Moreover, technical reports of different associations focused on renewable energies and fuels will be used to obtain data on the recent commercial production. Quantitative data regarding, among others, production cost and GHG emission of these reports will then be combined with the research articles obtained from databases. Search terms including 'biofuel production process', 'bioethanol/biodiesel/biomethane production', 'feedstock biofuels', 'biomass conversion', 'biofuel conversion technology', 'production cost biofuels', 'policies biofuel production', 'water use biofuel', 'life cycle analysis biofuel', 'politics biofuel', and 'land use biofuel', will be used to obtain the articles from the literature database.

From the analysis of the different feedstock types and conversion processes for bioethanol, biodiesel, and biomethane, a simplified overview of the production pathways is provided in Figure 1. Overall, the feedstocks can be subdivided in the first-generation complete crops and the second-generation, or advanced, biofuels from residues, wastes, and lignocellulosic biomass.

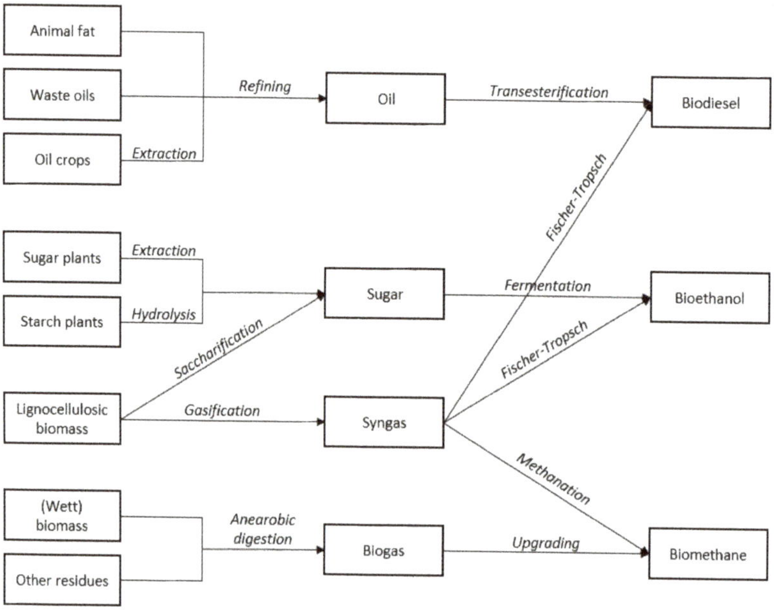

Figure 1. The different production pathways from feedstock to biofuel, self-constructed. The arrows indicate conversion processes, and the rectangles indicate the inputs and outputs of those conversion processes.

The factors from the sustainability facets that are directly affected by the biofuel industry will be determined by a PESTLE analysis. The final deliverable, the interrelation between the biofuel production characteristics, and the described sustainability factors, will be determined by an assessment of both qualitative and quantitative data obtained using the described literature research. In addition to the definite results, a discussion of the results will be incorporated to address and include the complexity of the biofuel industry and the linkages between the different sustainability facets.

2.2. PESTLE Analysis

The PESTLE analysis will be used to evaluate the external influences on the biofuel industry. It originates from marketing analysis, but these external influences on the industry will be used to determine the factors influencing the long-term sustainability of the industry. PESTLE stands for Political, Economic, Social, Technological, Legal, and Environmental. It provides a broad view of the complete environment of the biofuel industry. The analysis will determine the factors for the economic, environmental, socioecological, and geopolitical sustainability facets. It provides the framework for the correlation with the production technologies to determine the strength and weaknesses of the different production pathways.

2.2.1. Political and Legal

The political and legal aspects of the biofuel industry in the EU are driven by directives developed by the European Commission such as the RED. Each country has developed its own legal framework to support the national biofuel industry in achieving the target of 10% renewable transportation fuel by 2020 [12]. Recently, the RED has developed in recognizing the effect of indirect land-use change [33,34]. Furthermore, the competitiveness of biofuels in industrialized countries is driven by the subsidies and import tariffs in place. These tools create a significant barrier to international trade from the EU to other countries and therefore create a captive market for the national biofuel producers in the EU [103].

Politically, there are several factors impacting the national biofuel market, including employment laws and tax policies. However, to understand the overall political and legal impact on the EU biofuel market, the focus is on the differences in support policies for the biofuel development for each country. The support system for each biofuel significantly impacts the long-term sustainability of the biofuels.

2.2.2. Economical

As mentioned, the political aspect influences the economics of the biofuel industry by providing subsidies and creating import tariffs to protect and support the market. The biofuel industry, like any other industry, will be affected by changes in the economic environment which are, among others, tax, interest and exchange rates. However, specifically for the long-term sustainability of the biofuel market, the factor that will be addressed is the demand and supply of biofuel. The production cost of the different biofuels relates to the price, which affects the demand and therefore influences the supply. This circle is impacted by political influences, social behavior, and the development of technology. Hence, the application of PESTLE analysis to obtain a complete overview of the long-term sustainability of the biofuel industry.

As a replacer of fossil fuels, the competitiveness of biofuels does not only depend on its own production cost but even more so on the price of fossil fuels. Moreover, the current subsidies largely influence the competitiveness of the biofuel production cost. Biodiesel, bioethanol, and biomethane all have their own economic sustainability, influenced by factors including the feedstock, the type of conversion technologies, and the number of required processing steps. First, the cost of buying and cultivating feedstock differs depending on the type. Moreover, each feedstock has its own productivity influencing the overall cost per unit or volume of the feedstock [95]. Depending on the feedstock and the type of biofuel, there is a different conversion technology to produce the biofuel. Each conversion type has its own cost to implement and execute in a production process. However, the cost will develop with time, influenced by the learning and scaling effects [104]. In addition, the type of feedstock influences the number of production steps, since some types require pre-treatment or purification processes.

2.2.3. Technological

First, the technological aspect covers the technological developments and innovations in the industry [105–107]. The scope of this research on commercially available production technologies limits the effect of innovation on long-term sustainability. It is, however, important to notice that the development of the third generation biofuels, produced from algae, is a future development that could render the biofuels considered in this research relatively less sustainable [108–115].

Important to consider, is the overall development of the fuel and transportation industry, as electrical transportations vehicles become more important. The consideration of starting a biofuel production company at this moment of time should include the expected life cycle of the industry. If the electric vehicle industry developments increase significantly, the investments in biofuels could be unprofitable because of a decrease in demand. The technological influence on the biofuel industry highlights the importance of the expected demand and raises the question: Is the expected time until the demand of biofuel that will decrease due to other technologies long enough to be a profitable investment? On the other hand, the ability to convert the production process or use it for other applications will influence the expected lifetime of the investment.

2.2.4. Social

The influence of the social aspect of the biofuel industry relates to social changes and social stability. The common examples are the competition with the food supply and the cost of biofuels [95]. Moreover, water use in the biofuel industry will impact social stability. Water is already a scarce resource in many parts of the world, and the continuing development of the biofuel industry can add to the existing pressures [116].

A significant amount of the feedstock discussed in the previous chapter are also agricultural products used in food production, for example, corn and sugar beet. Therefore, the development of biofuel production has a significant impact on the world's agricultural market and food security [117]. Moreover, cultivation of the biofuel feedstock competes directly for land with other food crops such as coffee beans or rice. The feedstock that is also used for food and feed production including corn, wheat, sugar cane, soybean, rapeseed, and sunflowers are denoted as the first-generation feedstock. On the other hand, the second-generation feedstock, which includes lignocellulosic biomass and waste oils, do more increasingly support the co-existence of biofuel and food production [1].

Due to the existing pressure on water, the water consumption of the cultivation and conversion processes of the biofuels impact the overall sustainability of the production processes [116]. The different feedstocks will be analyzed on the use of water in the harvesting techniques. Moreover, the relative water consumption of the production process for each biofuel and feedstock will influence its impact on water scarcity. The ability to recycle the water streams in the process, thus reducing the net water use, is a significant factor that affects the performance of the process regarding water use.

2.2.5. Environmental

In the case of the biofuel industry, the environmental facet is extremely important since the reduction of the GHG emission is one of the key components in the sustainable development and one of the factors driving the transition to biofuels [118]. Moreover, it is important to note the interrelation between the social and the environmental aspects influencing the biofuel industry. This research defined the socioecological facet to summarize the factors that influence both aspects, including water scarcity and food competition. It also includes the aspect of land use for the cultivation of feedstock that has a large impact on the environment and social factors of the biofuel industry. Several life cycle assessments (LCA) obtained from literature, describe the improvement in GHG emission for different types of biofuels. Important are the different aspects influencing the GHG emission of biofuel production. The aspects include the type and management of the feedstock and the conversion technologies [119]. However, the results may also differ depending on the system boundaries of the LCA. Hence, the importance of an overall assessment of different sources in the following analysis. The general assumption in the LCA studies, regarding the GHG emission, is that CO_2 emissions from biomass combustion are climate-neutral due to the biomass absorption of CO_2 during growing [120]. Hence, the GHGs will depend more on the production processes surrounding biomass combustion.

The aspect of land-use incorporates the essential difference between the restoration of degraded farmlands or removing forests for biofuels [121]. The term indirect land-use change (ILUC) describes the change of natural environments to croplands to grow crops that replace the feedstock used for biofuels. Essentially, it is the effect of competing with the same resources as the food industry. The ILUC leads not only to a loss in biodiversity but also increases the GHG emission and impacts the prices of food [122]. The potential GHG savings are important for biofuel development, but they are reduced due to land-use change by, for example, deforestation. The exact land-use effects depend on the type of land that will be used and if deforestation is avoided [14]. These motives are hard to predict, therefore this study will take the land requirement of the feedstock, feedstock productivity, and the ability to grow on more degraded land as key factors influencing the land use [14]. The conversion of waste streams to biofuels is an important example of a feedstock that minimizes the land use. The interrelation between the land use and the environmental sustainability is recognized, but it is considered as socio-ecological due to its causal relationship with feedstock cultivation and its effect on the biodiversity, food competition, and water use.

Finally, biodiversity is also affected by the increase in crop cultivation due to biofuel development [123]. Biodiversity is supported by an environment with multiple crop species, while biofuel production yields are increased by the reduction of feedstock types that are cultivated in one area. The water use of the biofuel production can also be argued as an environmental aspect, hence the allocation to the socio-ecological facet.

2.2.6. Overview of Sustainability Framework

The results of the PESTLE analysis are summarized in Table 1. The collected factors will be correlated to the different production technologies, depending on the biofuel, feedstock, and conversion technology.

Table 1. The results of the PESTLE analysis as obtained are summarized per subcategory.

Political & Legal	Economic	Social	Technological	Environmental
• RED • National Policies	• Cost and efficiency • Feedstock production • Conversion process	• Price • Food competition • Water scarcity	High technological development of both biofuel industry and other renewable transportation industries	• GHG emissions • Feedstock and biofuel production • Direct and indirect land use • Biodiversity

The different factors from the PESTLE analysis all affect the long-term sustainability of the biofuel industry. The factors are subdivided into the sustainability facets where both the political and economic factors are directly assigned to the geopolitical and economic sustainability facet, respectively. Socioecological governance comprises the factors of food competition, water scarcity, and biodiversity, all affected by the discussed land-use change phenomena. The GHG emissions are covered in the environmental facet, which emphasizes one of the driving forces of biofuel development.

3. Correlation between the Production Processes and the Sustainability Framework

3.1. Economic Landscape

First, the distinction of biofuel produced from food crops and biofuel produced from waste and residues streams is often used in literature and is referred to as first- and second-generation biofuel, respectively. Regardless of the large variety in the production cost of biofuel that is influenced by feedstock, conversion process, the scale of the production, and the region, a general trend is determined. The first-generation biofuels have a significantly lower capital cost compared to the second generation. However, the feedstock cost of the first-generation biofuel poses a threat to its viability since it generally represents around 60–90% of the total production cost [35]. In the long term, the feedstock cost accounts for 30–45% of the total production cost for biofuels from lignocellulosic biomass. The second-generation biofuels will become competitive as the high capital costs reduce by the development of the relatively new conversion technologies. Nevertheless, the price of oil also impacts the biofuel industry to the extent that an oil price below US$80 per barrel will deem the second-generation biofuel uncompetitive with the fossil fuel market for the next 30 years [124].

Due to the high contribution of the feedstock to the total cost of the first-generation biofuel, the crop feedstocks are evaluated on, among other aspects, their efficiency. Table 2 indicates the high potential energy yield of the sugar feedstock compared to the starch feedstock. However, the conversion efficiency strictly dictates the overall productivity of the complete feedstock to the fuel process. It is notable to mention that raw sugarcane is not produced in the EU, it is always imported in the framework of preferential trading relationships. Raw sugarcane is processed in European countries contributing to the biofuel industry. Table 4 shows some results where the bioethanol from sugar crops still is superior considering the overall productivity. The energy productivities indicate a lower efficiency in terms of liters/ha for biodiesel crops production compared to both bioethanol and biomethane. However, the energy in terms of GJ shows more similar values, implying that the energy value of bioethanol in GJ/liters is significantly higher than that of biodiesel. The bioethanol from lignocellulosic biomass, see the cereal straw in Table 4 does not have competitive energy productivity. The overall economic sustainability could, however, profit from the low cost of waste streams as a feedstock [125]. Similar conditions apply to the residue feedstock-types used for both biochemical and thermochemical biodiesel. The woody crops and short rotation plants do show competitive energy productivity values of around 90 GJ/ha for thermochemically produced biodiesel.

Table 2. The yields of possible crop feedstock for biofuels in terms of its weight (tons) and energy (GJ) per hectare of cropland [126–128].

Feedstock	Yield (t/ha)	Yield (GJ/ha)
Sugar cane	68.7–70.9	370
Sugar beet	61.5–68.9	243.7–281
Maize	3.9–5.8	61
Wheat	3.5–7.9	77–119.4
Rapeseed	2.2–3.5	74–84.4
Sunflower seeds	1.8–2.4	58
Soya beans	2.8–2.9	56
Woody crops	11–27	87–239.4

In terms of availability, the cultivation of crops as a biofuel feedstock is not directly a problem as additional land could be acquired. It does significantly affect other sustainability facets of the industry as will be discussed later this chapter. The non-food crop feedstocks for biofuels including the woody residues used cooking oil, and animal fats also have an application in competing sectors. Nevertheless, it is determined that, as the total amount of these feedstock-types available is significantly larger than the demand from the competing sectors, the feedstock is available in a considerable amount for the biofuel industry. However, the availability of waste and residue streams has its limitations since the dedicate production of that core product for biofuels eliminates practically all benefits of the feedstock, which will be encountered when the demand increases.

Now, the efficiency of feedstock does influence economic competitiveness, but in the end, the cost of the energy source is the most important. Therefore, considering the cost not only with respect to volume but also the energy contribution is key to the sustainability assessment. The trend in first- and second-generation biofuels and the influence of the oil price, as described before, is also visible in Table 3, where the cost is normalized with respect to energy productivity. The values of energy productivity used in the previous study [125], from Table 3, differ from the values stated in Table 4, and these types of values are influenced, among others, by the region of feedstock origin and the weather [128]. However, the key message to extract is that for production of one fuel, the common values to consider are the cost per volume of fuel, whereas for the final consumption it is about the energy that can be extracted from the volume of fuel which will influence the cost assessment. Table 3 shows that the biodiesel produced from waste oil has a competitive production cost that will decrease even further with the development of the technology. The bioethanol from lignocellulosic biomass shows the prospects of the reduction in capital cost that directly creates a competitive production cost in the year 2020. The cost of conversion for the crop feedstock will not change significantly over time as the technology is matured, implying that the increase in production cost is imposed by the expected increase of crop prices [125]. Table 3 also depicts the smaller cost range for lignocellulosic bioethanol and biodiesel from waste oils, which implies that these products are the least sensitive for oil price changes. As the cost of the conversion process is more significant for the second-generation biofuel, the data in Table 5 addresses the different types of cost for both the biochemical and the thermochemical production pathway.

Table 3. The production cost for different biochemical produced biofuels and feedstock under varying oil price scenarios for both the year 2015 and 2020. The results are adapted from [125]. The total costs are normalized based on the energy density of the biofuel with respect to the energy density of fossil fuel to ensure the comparability of the total cost of the different biofuels.

(Bio-) Fuel	Raw Material	Crude Oil Price [€/Barrel]	Total Costs Normalized on Energy Density (€Cent/L)	
			2015	2020
Fossil fuel	Crude oil	50–200	36.5–130.8	36.5–130.8
Bioethanol	Maize	50–200	105.6–140.2	110.6–145.3
Bioethanol	Wheat	50–200	136.4–186.1	151.3–202.6
Bioethanol	Lignocellulosic waste material	50–200	157.3–171.2	81.5–95.4
Biodiesel	Rapeseed oil	50–200	117.5–171.4	138.4–192.3
Biodiesel	Palm oil	50–200	70.0–121.9	63.7–115.5
Biodiesel	Waste oil	50–200	61.8–89.9	45.4–73.6

Table 4. The yields of the different biofuel types per hectare of crop feedstock in terms of energy (GJ) and volume (liters) expected in the year 2020. The conversion technologies are all biochemical with exception from the thermochemical Fischer-Tropsch (FT) biodiesel. Adapted from [129].

Type of Biofuel	Feedstock	Energy Productivity 2020 (GJ/ha)	Energy Productivity 2020 (L/ha)
Bioethanol	Wheat	42	2000
	Maize	64	3030
	Barley	38	1770
	Sugar beet	145	6840
	Sugar cane	118	5570
Biomethane	Silage Maize	123	
Biodiesel	Sunflower oil	24.5	740
	Palm oil	88	2660
	Rapeseed	52	1570
	Soybean	17	530
Bioethanol	Cereal straw	15	710
FT Biodiesel	Miscanthus & Switchgrass	90	
	Short rotation plantation	97	
	Forest residues	5.8/year	

First, a large variation in feedstock price for the biochemical production of biomethane is apparent. The opportunity for biofuels is the low feedstock prices that can be obtained from using waste streams, which will be the type of zero cost feedstock [130]. The range of the production cost for biochemically produced biomethane can thus largely be allocated to the type of feedstock. Therefore, the more advanced and less mature thermochemical processes do show a slightly higher production cost compared to the biomethane after biogas upgrading. For the biochemical produced bioethanol from lignocellulosic biomass, the production cost will be higher as for crop-based bioethanol due to the additional process steps for lignocellulosic biofuels as discussed in the earlier chapter.

Table 5. Overview of the production cost (range) per unit of energy of both the biochemical (biomethane and bioethanol) and thermochemical (biomethane and biodiesel) for predefined feedstock. The cost of the biochemical conversion of biomethane covers all the different feedstock for anaerobic digestion. Adapted from [130].

Type of Biofuel	Conversion Process	Feedstock	Feedstock Price (€/MWh)	Production Cost Range (€/MWh)	Production Cost Range (€/GJ)
Biomethane	Biochemical	-	0–80	40–120	11–34
Bioethanol	Biochemical	Cellulosic biomass	10–13	85–103	24–29
Biomethane	Thermochemical	Waste and wood biomass	10–20	56–91	16–25
Biodiesel	Thermochemical	Wood biomass	10–20	90–139	25–35

For the overall economic sustainability of the different biofuels, a trend is clear that the expected feedstock cost favors the use of biomass, wastes, and residues. Only considering the feedstock cost, the fuels produced from lignocellulosic biomass: All thermochemical production process for the three biofuels, and the biochemical pathway for bioethanol, and the production from wastes and residues, which include the biochemical production of both biodiesel and biomethane. Hence, the biomethane has the highest number of production possibilities which provide a competitive feedstock cost. Moreover, the biochemical production of biomethane from biogas is a more mature technology, similar to the other biochemical production processes from crop feedstock that show competitive cost figures. However, the capital cost of the second-generation feedstock is expected to become competitive due to the development of the industry. A similar effect needs to apply to the thermochemical production path to support the economic sustainability of the biofuels produced via the syngas pathway from lignocellulosic biomass. To invest in these advanced biofuels in the current circumstance, long-term policy support is the only tool to provide certainty that investors will achieve a return on investment [131]. An extended scope, outside of the biofuel industry, shows that a low oil price could still render all the biofuels less economic sustainable due to the resulting uncompetitive production cost.

For the deployment of biomethane as fuel, the additional cost barrier exists of extending the distribution and fueling infrastructure of gaseous fuels [132]. Here the support, both in realizing and financing, of the national governments is necessary to allow a connection with the natural gas grid system. In addition to the investments required to start a production plant for one of the biofuels, these infrastructure costs for biomethane do affect its economic sustainability but they are essential to ensure demand for the fuel.

3.2. Environmental Governance

The GHG emissions will be correlated to the different types of feedstock and the different production processes. A variance in the GHG emission among the feedstock types is especially highlighted by the recent attention to the effect of indirect land-use change imposed by crop feedstock.

A broad overview of the emissions from the supply chain of the biofuels is depicted in Figure 2, distinguishing between the contribution of cultivation, processing, and transport to the total emission of the supply chain [128]. The study used the BioGrace [133] tool for calculations of the different types of emission that are expressed in CO_2-equivalents. The data in Figure 2 shows clearly that cultivation of the feedstock, if applicable, contributes on average for at least 50% to the total supply-chain emissions. Therefore, both biomethane production and biodiesel from waste oils have a superior position by non-agricultural waste streams as feedstock. The agricultural waste stream of lignocellulosic biomass used in the advanced ethanol and FT diesel contributes to the GHG effect by the utilization of fertilizers during cultivation [128].

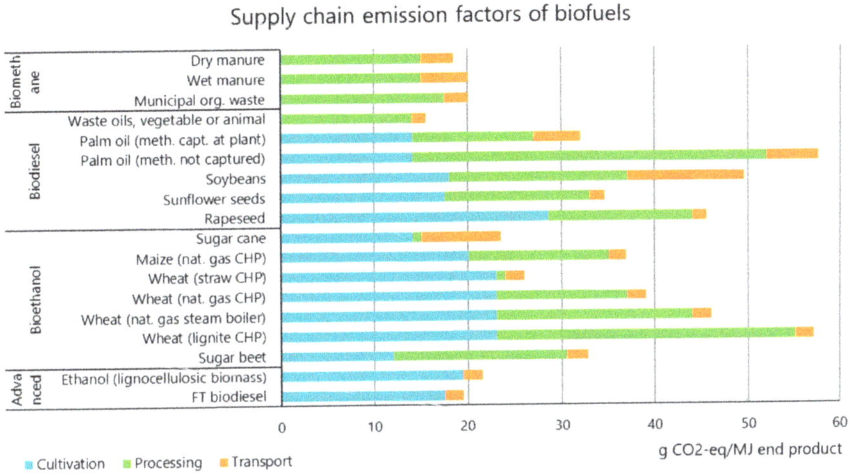

Figure 2. Supply chain emission factors expressed in the equivalent grams of CO_2/MJ of the end product, adapted from [128]. The results are based on [134] for the advanced biofuels and based on BioGrace [133] for the others, taking co-products into account. All processes are biochemical except for the thermochemical Fischer-Tropsch (FT) biodiesel. The text between parentheses in the ethanol category refers to the feedstock for processing.

The emissions during the conversion processes of the biofuels are influenced by the fossil energy consumption as clearly depicted in the case of lignocellulosic biomass where it is assumed that the conversion energy will be derived from the biomass itself. The same applies to the results depicted for sugar cane, where the conversion emission are calculated under the assumption that energy is produced from the by-product bagasse.

Overall, this discussion of recycling of waste streams in the biofuel processes creates not only the challenge in the analysis of emissions but will also affect production costs. Another example is the case where bagasse can also be used as lignocellulosic biomass for biochemically produced bioethanol or the thermochemical production. However, in this case, the bagasse can be a waste stream of the food industry compared to an internal waste stream as with the utilization of sugar cane. Another possibility one could imagine is the replacement of fossil fuels used in transport and conversion process by biofuels.

For environmental sustainability purely, this implementation would be ideal since it will reduce the emissions of the supply chain. However, considering the economic landscape, it would only weaken the efficiency of a production process since the land use will increase to produce similar amounts of biofuel. Such implementation would thus require the development of policies and production criteria and to a higher extent analysis of emissions imposed by the additional land use. If a fossil system is an alternative to using residues for energy production, the use of food waste for biofuel production can show overall negative GHG emission effect since the loss of energy will be compensated with fossil fuels. This relation will develop on the long-term as sources for the electricity and heat sector will be less pollutive for the environment [135].

Overall the land-use change imposes reduces the environmental sustainability of biofuels produced from food crops compared to alternatives. There is even research that reports a net increase in GHG emission for food crop biofuels imposed by the land use impacts compared to fossil fuels [135]. Table 6 clearly shows the higher land-use emissions for the biodiesel feedstock compared to bioethanol. First of all, a relation is visible between crop productivity and the land-use emission, as less land is required to produce the same amount of energy. This effect benefits the first-generation bioethanol feedstock over the oil crops for biodiesel. Moreover, the land used to grow the crops impacts the overall emission. For biodiesel, specifically for palm oil, the emission is a consequence of the high conversion of peatlands into palm cultivation areas [129]. This impact is confirmed by the study of [128], where direct land-use change from grassland is significantly favorable compared to the conversion of forest lands to biofuel croplands. Hence, the negative values for the fast-growing plantation for FT biodiesel, which are particularly grown on marginal lands and thus improve the conditions of the land resulting in an overall reduction of emissions. The land-use emission for biomethane production shows no significant differences with the bioethanol production from crops. The use of agricultural wastes from the food industry could to an extent reduce these emissions since the cultivation is not directly influenced by the biofuel demand.

Table 6. The land use emissions for various feedstock that include both the direct and indirect effects of both land use and land-use change. All conversion processes are biochemical, except the thermochemically produced Fischer-Tropsch (FT) biodiesel. Adapted from [129].

Type of Biofuel	Feedstock	Land Use Emission (g CO_2/MJ)
Bioethanol	Wheat	34
	Maize	14
	Barley	36
	Sugar beet	15
	Sugar cane	17
Biomethane	Silage Maize	21
Biodiesel	Sunflower oil	63
	Palm oil	231
	Rapeseed	65
	Soybean	150
Bioethanol	Cereal straw	16
FT Biodiesel	Miscanthus & Switchgrass	−12
	Short rotation plantation	−29

An analysis of recent studies regarding the ILUC emissions shows varying results and the overall ILUC effects are therefore considered difficult to precisely define [136]. However, wood and crop residues are considered promising as the ILUC GHG emission are considerably lower than the first-generation crop-based biofuels [129,137]. The study of [136] does indicate a trend where on average biodiesel has higher ILUC emission (median of 52 g CO_2-eq/MJ) compared to first-generation ethanol (median of 21 g CO_2-eq/MJ), with the sugar crops showing the lowest ILUC emissions.

The negative effect of the application of crops as a biofuel feedstock is once more highlighted by the land-use emission analysis. The general emission from the biofuel supply chain already did not depict favorable equivalent CO_2 factors. Overall, the environmental sustainability of crop feedstocks is impacted especially with the available superior feedstock. Not only the supply chain emissions are competitive, but the residue streams as feedstock also do not impose the land-use impacts seen by the first-generation bioethanol and biodiesel feedstock. In addition to the thermochemical conversion of grasses to biodiesel that shows negative land-use emission, the biochemical conversion of waste oils already shows an environmental favorable supply chain and the land-use effects are not applicable.

3.3. *Socioecological Aspects*

The essence of the socioecological sustainability originates from the origin of the feedstock, whether it is cultivated on agricultural land or a waste stream from an industrial process for example. Considering only first-generation bioethanol and biodiesel, the study of [138] analyzed the water, land, and food use with data from 85% of the global bioethanol consumption and 81% of the global biodiesel consumption in 2013. The results indicate that for only crop feedstock, biodiesel requires on average 90,000 m^3/TJ of water compared to 74,000 m^3/TJ for bioethanol. The land required is on average 9 ha/TJ for bioethanol and 29 ha/TJ for biodiesel. However, based on the food calories used for the production of the first-generation biofuels, the biodiesel is less competitive with the food industry with an average of 95 people/TJ of biodiesel that could be fed compared to 107 people/TJ of bioethanol. Considering the production processes of both bioethanol and biodiesel it can thus be concluded that the overall efficiency of land to biofuel is significantly lower for biodiesel. However, the use of waste oils, for example, the waste cooking oils from households, does not require the extraction processes compared to the oil crops, which would favor the increase the production yields per volume of feedstock used. In comparison, the biochemical production of bioethanol from lignocellulosic biomass still does not convert all biomass to biofuel. The thermochemical conversion technologies would reduce the waste stream since the complete biomass can be converted to syngas. However, this does not directly implicate that thermochemical conversion can produce more biofuel from less land since not all the feedstock has the same amount of bioethanol yield. Moreover, the biochemical production processes do consider the recycling of waste streams such as the use of lignin to generate the required heat for the process, as discussed in the conversion technologies. Overall, the distinction between food and non-food feedstock is essential for the discussion and implication of land use, as is recognized by the European Commission, which presented the Indirect Land Use Change (ILUC) Directive in 2015 to tackle those negative effects of food-based feedstock. It limits the share of biofuels from crops grown on agricultural land to 7% and directs the Member States of the EU to develop national targets for the second generation biofuels [34].

Considering the feedstock for biomethane, the biofuel has a strong socioecological position due to the extensive possibilities in the application of feedstock other than complete agricultural crops. However, bioethanol could still be a competitive biofuel as national targets are set to develop the production from lignocellulosic biomass. On the other hand, biodiesel production from used cooking oils is even more technologically mature, as it is the feedstock for 86% of the UK biodiesel [139]. Thermochemical conversion to syngas implies that the same feedstock can be used for all three biofuels, implying that socioecological sustainability is less a decisive factor for the biofuel choice with this technology. Since the feedstock for the gasification process originates mainly from residues and biomass, the impact of thermochemical production on the socioecological sustainability factors will generally be less as to the biochemical conversion pathway, especially with the food crop feedstock. However, conversion efficiencies of the different technology could reduce this effect and would increase the possibilities for feedstock that impose less pressure on agricultural land use.

To reduce the impact of land use and the competition with the food industry there are crops that are non-edible and grow well on marginal lands including the discussed Jatropha (Bhuiya et al., 2016). However, similar to the other crops, these crops do also require significant amounts of water to grow and cultivate compared to the water footprint of crop residues [140]. Table 7 is indicative of the reduction in water use if residues of crops are used for biofuel production instead of complete crops. It would only be favorable if the residues could be obtained from normal agricultural activities, such that it is a waste stream of another industry instead of specially cultivated to produce biofuel. Further analysis of the crop yields in Table 7 shows that the water footprint is lowest for the sugar feedstock used for bioethanol and that the oil crops for biodiesel have a higher water footprint. This is the same trend as depicted by the first-generation biofuel data of [138] as discussed.

The change in biodiversity due to the biofuel industry is mainly driven by land use [141]. The negative impact of deforestation on biodiversity has already been described. However, for the conversion of abandoned cropland or marginal lands to biofuel croplands, the impacts are not that clear. In general, the use of the second-generation crops is considered to have a less negative influence on the biodiversity compared to the first-generation [141]. The improvements in the biodiversity impacts by the cultivation of second-generation feedstock are also supported by the study of [142].

Table 7. The water footprint in m^3 per ton of crop yields and residues. It states the amount of water used in growing the useful part of the crop (the yield), and for what is considered the residue [140].

Crop	Water Footprint of Crop Yield (m^3/t)	Water Footprint Residue (m^3/t)
Sugar cane	176	72
Corn	961	205
Rice	1523	129
Wheat	1633	140
Sugar beet	103	47
Cassava	476	87
Soybean	2002	188
Rapeseed	1583	205
Cotton	3796	154
Sunflower	2014	636

For the food competition and water use issues, the important driver is the type of feedstock and the land that cultivation requires. Biodiversity does also improve by minimization of crop feedstock but is to a higher extent influenced by the type of land that is converted for cultivation [142]. In other words, biodiversity is the easier factor of the socioecological facet to improve by creating legislation that restricts certain areas of land to be used for cultivation of biofuel crops.

3.4. Geopolitical Substratum

Towards the renewable energy goals of the EU of the year 2020, all 28 member states are obliged to report on their progress every two years. The binding target for the transportation sector is to obtain a renewable energy share of at least 10%, which support the development of the biofuel industry to achieve this target [12,33]. As the biofuels are developed as renewable energies, the environmental and socioecological facets can be considered as drivers for the biofuel policies. Therefore, the geopolitical sustainability, the extent to which distinctive parts of the biofuel industry are supported by the geographic characteristics and the developed policies, is strongly affected by the assessment of its environmental and socioecological sustainability. A substantial amount of the current policies now incorporates a system of certification that ensures sustainable production of biofuels. To which extent the term 'sustainable' includes all facets covered in this research is difficult to determine, but the programs were initiated after the publication of reports about the interrelation of biofuel production and food shortages, biodiversity loss, and the land-use change [143]. Moreover, to directly reduce the impact of food crops on these types of issues, a restriction is introduced which limits the use of food crops, the sugar, starch, and oil crops, to 7% of all the feedstock used to produce biofuel.

Among the EU member states, Germany is the country with the highest level of development of bioenergy. Their target for the transportation sector is set at 13.8% renewable energy by 2020 [144], aiming above the guideline of the EU. In comparison, Poland has aligned the 2020 target of renewable energy in transport with the minimum 10% imposed by the EU but aiming for 8.5% renewable fuels and 1.5% contribution from electricity used in transport [145].

The Netherlands was the first, in 2009, EU country to have legislation in place that supported the production of biofuels from waste, residues, and lignocellulosic biomass by counting these advanced biofuels double towards the targets set for the industry [146]. In practice, these types of biofuel contributed for 50% to the total renewable energy used in the Netherlands in 2016, of which used cooking oil and animal fats for biodiesel production were the primary sources. Recently, Italy as a pioneer legally required all fuel suppliers to include 0.6% advanced biofuels in the gasoline and diesel produced from 2018 [147]. In contrast, Poland did not implement the directive supporting the advanced biofuel in its legal system, including a double counting system, until 2018 [145]. Hence, no data on the use of this type of feedstock in biofuel production is available in the prior years. Romania, another country in non-western Europe, experiences similar problems in supporting advanced biofuels. In this case, they explicitly refer to budgetary problems that limit their development of support schemes especially required for the infrastructure that is needed to produce the more costly biofuels from wastes, residues, and lignocellulosic biomass [148].

Overall in the EU, there is a direct intention to limit the use of first-generation biofuels. The alternatives, including the use of waste and residues, do require additional support that cannot yet be provided by all countries. Hence, the geopolitical sustainability of food crop-based biofuels decreases, while the use of advanced biofuels is increasingly promoted. However, in the current state of policies, the economic support for advanced biofuels seems more developed in western Europe, whereas other countries face difficulties in providing a support system that can lower the higher capital cost of advanced biofuel production.

The development of the biofuel industry can be linked to the availability of feedstock as input to the production process. From Figures 3 and 4, it can be derived that, especially for bioethanol, a large majority of the feedstock used for the EU biofuel industry originate from Europe itself. Regarding the support of the biofuel industry by the development of policies, the application of native feedstock will support policies that address the whole supply chain of the biofuel industry. However, there are a growing number of voluntary schemes that can be used worldwide to prove compliance with the EU biofuel sustainability criteria. The 63.7% of biodiesel produced with feedstock from the EU originates for 33% from used cooking oil and animal fats [149]. These feedstock types are available worldwide and meet the trend of requirements set for the sustainability of biofuel feedstock. The 20.6% of feedstock imported from Malaysia and Indonesia, which is primarily palm oil, is not compliant with the developing policies. Recent criteria set by the EU deem all the palm oil produced on land larger than 2 ha unsustainable, which implies that biofuel from palm oil cannot be counted towards the renewable energy targets imposed by the EU [150].

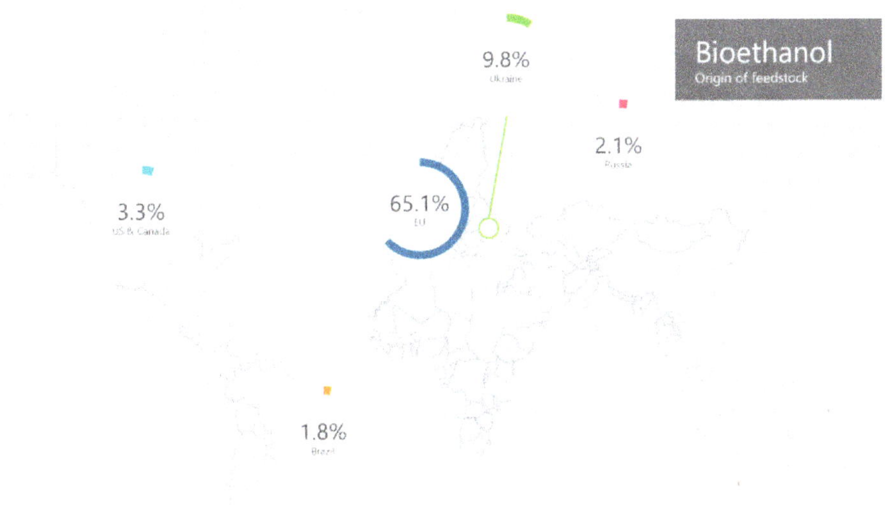

Figure 3. The origin of the feedstock used for bioethanol consumed in the EU in 2016. The percentages with respect to the total bioethanol consumption are included. The figure is adapted from data of [149].

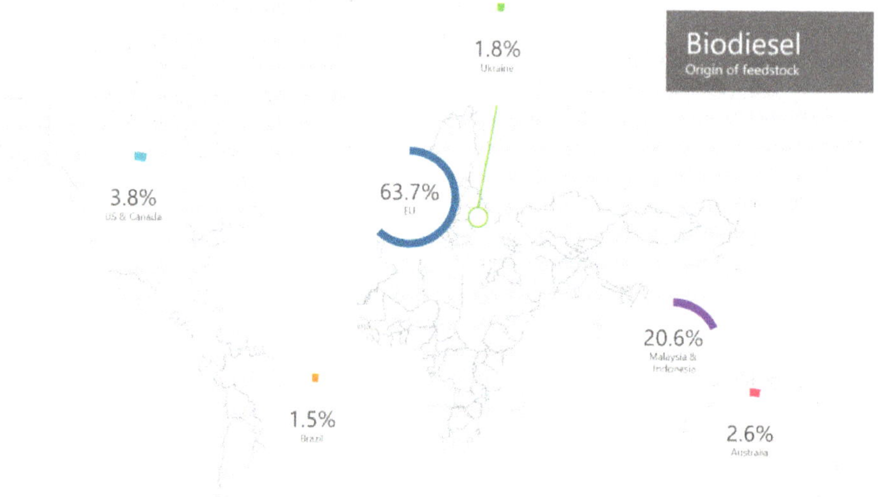

Figure 4. The origin of the feedstock used for biodiesel consumed in the EU in 2016. The percentages with respect to the total biodiesel consumption are included. The figure is adapted from data of [149].

The creation of the national laws to achieve the targets of EU is empowered in the RED, on the only condition that there is no legislation that restricts the free movement of fuels and feedstocks among the EU countries [150]. For countries outside of the EU, anti-dumping duties are imposed on, for example, the US ethanol in 2013, since that market price was below what EU producers could offer [151]. Hence, the consensus among the EU countries appears to be the use of the EU biofuel market to support the development of the EU production companies. This trend, together with the extent of imposed EU tariffs on biofuel, drives the EU towards the creation of an isolated biofuel market [151]. Depending on the trade of the feedstock, such an isolated market will have more impact on the variety of feedstocks used in the EU biofuel than on the distribution of the biofuel types that are consumed, since, for all biofuels, the necessary feedstock is available in the EU itself.

4. Discussion

From the start of this integration project, a broad research-oriented approach is deemed suited to address the existing knowledge gap that restricted biofuel development. The variety of the production characteristics and sustainability factors that are analyzed as a result of this approach highlight the complexity of the industry faced by policymakers and production companies. The research structures this complexity by identifying key relationships between the industry and its sustainability. However, some relationships are more extensively discussed and highlighted compared to others due to the lack of recent representative research. Further research could, for example, specifically address production characteristics of the biogas upgrading for the biomethane production to allow a more equal comparison with the more mature bioethanol and biodiesel processes. Moreover, further research should build on the positive relationships between a production characteristic and the sustainability identified in this research by analyzing how to optimize the positive effect and neutralize additional consequences on other sustainability facets. That knowledge will contribute to effective future development of biofuels by companies and governments while ensuring its sustainability.

An understanding of the key aspects of the biofuel production process and how they all affect the sustainability of the industry can contribute to the sustainable development of biofuels. To address the existing knowledge gap, this research correlates key production characteristics of bioethanol, biodiesel, and biomethane with the sustainability factors that are affected by the biofuel industry. The results show a significant number of possibilities in the supply chain of the biofuel industry. First, the feedstock for bioethanol and biodiesel can either be food crops, the first-generation, or lignocellulosic biomass, waste and residue streams, the second-generation. For biomethane, the main feedstock sources are waste and by-products from the agricultural, industrial, and municipal sectors. Second, the conversion process of feedstock to biofuel can be subdivided in the biochemical and the thermochemical pathway. The latter is commercially available as the syngas-route, where lignocellulosic biomass is the most common renewable feedstock.

The correlation executed in the research shows that the current technological developments in the industry significantly affect the competitiveness between different production characteristics with respect to economic sustainability. The characteristics of biomethane show a good potential to maintain a competitive production cost both in the current perspective as in the near future. The low feedstock cost of the second-generation biofuels gives a competitive advantage to the lignocellulosic bioethanol and biodiesel from waste oils via the biochemical production pathway which will only increase over time as the technologies become more mature. In general, the biofuels from lignocellulosic biomass produced via the thermochemical pathway become also more competitive as the capital cost will decrease. Overall, the economic sustainability of biodiesel and bioethanol from crop feedstock will decrease as the alternatives and their conversion technology develop.

The environmental impacts of the different biofuels differ significantly. Biomethane and biodiesel from waste oils show favorable properties because there are no emissions from crop cultivation and the land-use impacts are minimized. The ability to use the feedstock to generate energy contributes to a reduction in emission for some crops, including sugar cane and wheat, and the biofuels produced from lignocellulosic biomass. The latter also avoids largely all impacts from land use and land-use change. Additionally, the impact of land use and land-use change induced by crop feedstock affects socioecological sustainability. The use of residues and waste streams would reduce the impact on food competition, water use, and biodiversity. Biomethane shows the largest number of possibilities in avoiding crop feedstock and using different waste and residue stream. For the use of food crops, the environmental and socio-ecological sustainability of the bioethanol production process shows more favorable results compared to biodiesel production.

The drive for the biofuel production still originates from EU directives and the related government legislation. In policy development, there is a focus on the feedstock types and their origin, with limitations for the use of crop feedstock and the first initiatives being introduced that support the second-generation biofuels. However, the development is not uniform for all the EU countries, as not all countries have the economic capacity and infrastructure required for such extensive support. Overall, the results of the correlation can be used as the starting point for a multiple-scenario assessment which is required to develop a policy that supports the sustainable development of biofuels. Moreover, production companies can start an assessment of their production characteristics to ensure that these will suit future policies incorporating the importance of sustainable development.

5. Summary

A broad research-oriented approach was deemed suitable to address the existing knowledge gap that restricts biofuel development. The variety of the production characteristics and sustainability factors, that were analyzed as a result of this approach, highlight the complexity of the industry faced by policymakers and production companies. The research structured this complexity by identifying key relationships between the industry and its sustainability. However, some relationships were more extensively discussed and highlighted compared to others due to the lack of recent representative research. Further research could, for example, specifically address production characteristics of the biogas upgrading for the biomethane production to allow a more equal comparison with the more mature bioethanol and biodiesel processes. Moreover, further research should be built on the positive relationships between a production characteristic and the sustainability identified in this research by analysing how to optimize the positive effect and neutralise additional consequences on other sustainability facets. That knowledge will contribute to an effective future development of biofuels by companies and governments while ensuring its sustainability.

Author Contributions: Conceptualization, S.A.; methodology, S.A.; writing—original draft preparation, J.H. and S.A.; writing—review and editing, V.A. and G.J.W.E.

Funding: This research received no external funding.

Conflicts of Interest: The authors declare no conflict of interest.

References

1. Cook, P.J. CCS research development and deployment in a clean energy future: Lessons from Australia over the past two decades. *Engineering* **2017**, *3*, 477–484. [CrossRef]
2. Hoque, K.M.R.; Ozansoy, C.; Fahrioglu, M. Climate and size correction in European Union's Waste Framework Directive and R1 energy efficiency criteria. *Waste Manag. Res.* **2018**, *36*, 670–688. [CrossRef] [PubMed]
3. Longinos, S.N.; Bülbül, S.; Parlaktuna, M. Potential effects of methane hydrates to the environment. In Proceedings of the 12th Panhellenic Scientific Conference in Chemical Engineering, Athens, Greece, 29–31 May 2019.
4. Merey, S.; Longinos, S.N. The gas hydrate potential of the Eastern Mediterranean basin. *Bull. Min. Res. Exp.* **2019**, *159*, 1–10. [CrossRef]
5. Merey, S.; Longinos, S.N. The role of natural gas hydrate during natural gas transportation. *OHU J. Eng. Sci.* **2018**, *7*, 937–953. [CrossRef]
6. Longinos, S.N.; Merey, S. CO_2 sequestration in the mature oil fields: A case study of Bati Raman oil field, Turkey. In Proceedings of the International Conference on Energy Research, Alanya, Turkey, 31 October–2 November 2018.
7. Longinos, S.N. *Potential Environmental Challenges for Gas Hydrates*, 1st ed.; Lap Lambert Academic Publishing: Riga, Latvia, 2019; pp. 1–60.
8. Davis, L.A. The shale oil and gas revolution. *Engineering* **2018**, *4*, 438–439. [CrossRef]
9. Ranieri, L.; Mossa, G.; Pellegrino, R.; Digiesi, S. Energy recovery from the organic fraction of Municipal Solid Waste: A real options-based facility assessment. *Sustainability* **2018**, *10*, 368. [CrossRef]
10. Gowreesunker, B.L.; Tassou, S.; Atuonwu, J. Cost-energy optimum pathway for the UK food manufacturing industry to meet the UK national emission targets. *Energies* **2018**, *11*, 2630. [CrossRef]
11. Ajanovic, A. Biofuels versus food production: Does biofuels production increase food prices? *Energy* **2011**, *36*, 2070–2076. [CrossRef]
12. European Commission. Biofuels. 2019. Available online: https://ec.europa.eu/energy/en/topics/renewable-energy/biofuels (accessed on 15 May 2019).
13. European Commission. Renewable Energy Directive. 2019. Available online: https://ec.europa.eu/energy/en/topics/renewable-energy/renewable-energy-directive (accessed on 26 March 2019).
14. Linares, P.; Pérez-Arriaga, I.J. A sustainable framework for biofuels in Europe. *Energy Policy* **2013**, *52*, 166–169. [CrossRef]
15. Zhang, Z.; Han, W.; Chen, X.; Yang, N.; Lu, C.; Wang, Y. The Life-Cycle Environmental impact of recycling of restaurant food waste in Lanzhou, China. *Appl. Sci.* **2019**, *9*, 3608. [CrossRef]
16. Domínguez, J.; Gómez-Brandón, M.; Martínez-Cordeiro, H.; Lores, M. Bioconversion of Scotch broom into a high-quality organic fertiliser: Vermicomposting as a sustainable option. *Waste Manag. Res.* **2018**, *36*, 1092–1099. [CrossRef]
17. Trakulvichean, S.; Chaiprasert, P.; Otmakhova, J.; Songkasiri, W. Comparison of fermented animal feed and mushroom growth media as two value-added options for waste Cassava pulp management. *Waste Manag. Res.* **2017**, *35*, 1210–1219. [CrossRef] [PubMed]
18. Longjan, G.G.; Dehouche, Z. Nutrient characterisation and bioenergy potential of common Nigerian food wastes. *Waste Manag. Res.* **2018**, *36*, 426–435. [CrossRef] [PubMed]
19. Noya, I.; Inglezakis, V.; González-García, S.; Katsou, E.; Feijoo, G.; Moreira, M. Comparative environmental assessment of alternative waste management strategies in developing regions: A case study in Kazakhstan. *Waste Manag. Res.* **2018**, *36*, 689–697. [CrossRef] [PubMed]

20. Di Maria, F.; Sisani, F. Effectiveness of municipal solid waste incinerators in replacing other fuels. A primary energy balance approach for the EU28. *Waste Manag. Res.* **2018**, *36*, 942–951. [CrossRef] [PubMed]
21. Muradin, M.; Joachimiak-Lechman, K.; Foltynowicz, Z. Evaluation of eco-efficiency of two alternative agricultural biogas plants. *Appl. Sci.* **2018**, *8*, 2083. [CrossRef]
22. Charalampidis, I.; Karkatsoulis, P.; Capros, P. A regional economy-energy-transport model of the EU for assessing decarbonization in transport. *Energies* **2019**, *12*, 3128. [CrossRef]
23. Ribeiro, E.M.; Mambeli Barros, R.; Tiago Filho, G.L.; dos Santos, I.F.S.; Sampaio, L.C.; dos Santos, T.V.; de Freitas, J.V.R. Feasibility of biogas and energy generation from poultry manure in Brazil. *Waste Manag. Res.* **2018**, *36*, 221–235. [CrossRef]
24. Philippidis, G.; Bartelings, H.; Helming, J.; M'barek, R.; Smeets, E.; Van Meijl, H. The good, the bad and the uncertain: Bioenergy use in the European Union. *Energies* **2018**, *11*, 2703. [CrossRef]
25. Lauer, M.; Thrän, D. Flexible biogas in future energy systems—Sleeping beauty for a cheaper power generation. *Energies* **2018**, *11*, 761. [CrossRef]
26. Ferreira, E.T.D.F.; Balestieri, J.A.P. Comparative analysis of waste-to-energy alternatives for a low-capacity power plant in Brazil. *Waste Manag. Res.* **2018**, *36*, 247–258. [CrossRef] [PubMed]
27. Manni, M.; Coccia, V.; Cavalaglio, G.; Nicolini, A.; Petrozzi, A. Best practices for recovering rural abandoned towers through the installation of small-scale biogas plants. *Energies* **2017**, *10*, 1224. [CrossRef]
28. Park, J.K.; Chong, Y.G.; Tameda, K.; Lee, N.H. Methods for determining the methane generation potential and methane generation rate constant for the FOD model: a review. *Waste Manag. Res.* **2018**, *36*, 200–220. [CrossRef] [PubMed]
29. Achinas, S.; Achinas, V.; Euverink, G.J.W. A technological overview of biogas production from biowaste. *Engineering* **2017**, *3*, 299–307. [CrossRef]
30. Lindkvist, E.; Johansson, M.T.; Rosenqvist, J. Methodology for analysing energy demand in biogas production plants—A comparative study of two biogas plants. *Energies* **2017**, *10*, 1822. [CrossRef]
31. Safar, K.M.; Bux, M.R.; Aslam, U.M.; Shankar, B.A.; Goel, R.K. The feasibility of putrescible components of municipal solid waste for biomethane production at Hyderabad, Pakistan. *Waste Manag. Res.* **2018**, *36*, 169–182. [CrossRef] [PubMed]
32. EurObserv'ER. The state of renewable energies in Europe. In *18th EurObserv'ER Report*; EurObserv'ER: Paris, France, 2018.
33. European Commission. Renewable Energy: Progress Reports. 2019. Available online: https://ec.europa.eu/energy/en/topics/renewable-energy/progress-reports (accessed on 27 May 2019).
34. European Environment Agency. *Renewable Energy in Europe—2018: Recent Growth and Knock-on Effects*; European Environment Agency: Copenhagen, Denmark, 2018.
35. Royal Academy of Engineering. *Sustainability of Liquid Biofuels*; Royal Academy of Engineering: London, UK, 2017.
36. Moeller, L.; Bauer, A.; Wedwitschka, H.; Stinner, W.; Zehnsdorf, A. Crop characteristics of aquatic macrophytes for use as a substrate in anaerobic digestion plants—A study from Germany. *Energies* **2018**, *11*, 3016. [CrossRef]
37. Reißmann, D.; Thrän, D.; Bezama, A. How to identify suitable ways for the hydrothermal treatment of wet bio-waste? A critical review and methods proposal. *Waste Manag. Res.* **2018**, *36*, 912–923. [CrossRef]
38. Chatzikonstantinou, D.; Tremouli, A.; Papadopoulou, K.; Kanellos, G.; Lampropoulos, I.; Lyberatos, G. Bioelectricity production from fermentable household waste in a dual-chamber microbial fuel cell. *Waste Manag. Res.* **2018**, *36*, 1037–1042. [CrossRef]
39. Benato, A.; Macor, A. biogas engine waste heat recovery using organic rankine cycle. *Energies* **2017**, *10*, 327. [CrossRef]
40. Li, J.; Huang, H.; Huhetaoli; Osaka, Y.; Bai, Y.; Kobayashi, N.; Chen, Y. Combustion and heat release characteristics of biogas under hydrogen- and oxygen-enriched condition. *Energies* **2017**, *10*, 1200. [CrossRef]
41. Rosero-Henao, J.C.; Bueno, B.E.; de Souza, R.; Ribeiro, R.; Lopes de Oliveira, A.; Gomide, C.A.; Gomes, T.M.; Tommaso, G. Potential benefits of near critical and supercritical pre-treatment of lignocellulosic biomass towards anaerobic digestion. *Waste Manag. Res.* **2019**, *37*, 74–82. [CrossRef] [PubMed]
42. Oreggioni, G.D.; Gowreesunker, L.B.; Tassou, S.A.; Bianchi, G.; Reilly, M.; Kirby, M.E.; Toop, T.A.; Theodorou, M.K. Potential for energy production from farm wastes using anaerobic digestion in the UK: An economic comparison of different size plants. *Energies* **2017**, *10*, 1396. [CrossRef]

43. Milledge, J.J.; Nielsen, B.V.; Sadek, M.S.; Harvey, P.J. Effect of freshwater washing pretreatment on *Sargassum muticum* as a feedstock for biogas production. *Energies* **2018**, *11*, 1771. [CrossRef]
44. Carlini, M.; Mosconi, E.M.; Castellucci, S.; Villarini, M.; Colantoni, A. An economical evaluation of anaerobic digestion plants fed with organic agro-industrial waste. *Energies* **2017**, *10*, 1165. [CrossRef]
45. Bezergianni, S.; Dimitriadis, A.; Faussone, G.-C.; Karonis, D. Alternative diesel from waste plastics. *Energies* **2017**, *10*, 1750. [CrossRef]
46. Zhou, Y.; Peng, S.; Huang, X.; Wu, C.; Zhang, J. LNG–air mixture as a supplementary energy injection into a biogas distribution Network. *Energies* **2017**, *10*, 1902. [CrossRef]
47. Koçer, A.T.; Özçimen, D. Investigation of the biogas production potential from algal wastes. *Waste Manag. Res.* **2018**, *36*, 1100–1105. [CrossRef]
48. Achinas, S.; Euverink, G.J.W. Feasibility study of biogas production from hardly degradable material in co-inoculated bioreactor. *Energies* **2019**, *12*, 1040. [CrossRef]
49. Ghanimeh, S.; Khalil, C.A.; Ibrahim, E. Anaerobic digestion of food waste with aerobic post-treatment: Effect of fruit and vegetable content. *Waste Manag. Res.* **2018**, *36*, 965–974. [CrossRef]
50. Zhou, Z.; Tang, Y.; Chi, Y.; Ni, M.; Buekens, A. Waste-to-energy: A review of life cycle assessment and its extension methods. *Waste Manag. Res.* **2018**, *36*, 3–16. [CrossRef] [PubMed]
51. Carnevale, E.; Molari, G.; Vittuari, M. Used cooking oils in the biogas chain: a technical and economic assessment. *Energies* **2017**, *10*, 192. [CrossRef]
52. Achinas, S.; Li, Y.; Achinas, V.; Euverink, G.J.W. Biogas potential from the anaerobic digestion of potato peels: Process performance and kinetics evaluation. *Energies* **2019**, *12*, 2311. [CrossRef]
53. Kim, S.; Sung, T.; Kim, K.C. Thermodynamic performance analysis of a biogas-fuelled micro-gas turbine with a bottoming organic rankine cycle for sewage sludge and food waste treatment plants. *Energies* **2017**, *10*, 275. [CrossRef]
54. Calabrò, P.; Catalán, E.; Folino, A.; Sánchez, A.; Komilis, D. Effect of three pretreatment techniques on the chemical composition and on the methane yields of Opuntia ficus-indica (prickly pear) biomass. *Waste Manag. Res.* **2018**, *36*, 17–29. [CrossRef]
55. Achinas, S.; Euverink, G.J.W. Elevated biogas production from the anaerobic co-digestion of farmhouse waste: Insight into the process performance and kinetics. *Waste Manag. Res.* **2019**, in press. [CrossRef]
56. Sahajwalla, V. Green processes: Transforming waste into valuable resources. *Engineering* **2018**, *4*, 309–310. [CrossRef]
57. Chen, P.; Anderson, E.; Addy, M.; Zhang, R.; Cheng, Y.; Peng, P.; Ma, Y.; Fan, L.; Zhang, Y.; Lu, Q.; et al. Breakthrough technologies for the biorefining of organic solid and liquid wastes. *Engineering* **2018**, *4*, 574–580. [CrossRef]
58. RedCorn, R.; Fatemi, S.; Engelberth, A.S. Comparing end-use potential for industrial food-waste sources. *Engineering* **2018**, *4*, 371–380. [CrossRef]
59. Baute, K.; Van Eerd, L.L.; Robinson, D.E.; Sikkema, P.H.; Mushtaq, M.; Gilroyed, B.H. Comparing the biomass yield and biogas potential of *Phragmites australis* with *Miscanthus × giganteus* and *Panicum virgatum* Grown in Canada. *Energies* **2018**, *11*, 2198. [CrossRef]
60. Matsakas, L.; Gao, Q.; Jansson, S.; Rova, U.; Christakopoulos, P. Green conversion of municipal solid wastes into fuels and chemicals. *Electron. J. Biotechnol.* **2017**, *26*, 69–83. [CrossRef]
61. Di Matteo, U.; Nastasi, B.; Albo, A.; Astiaso Garcia, D. energy contribution of OFMSW (Organic Fraction of Municipal Solid Waste) to energy-environmental sustainability in urban areas at small scale. *Energies* **2017**, *10*, 229. [CrossRef]
62. Kim, S.; Sung, T.; Kim, K.C. performance and greenhouse gas reduction analysis of biogas-fueled solid-oxide fuel cells for a sewage sludge and food waste treatment facility. *Energies* **2018**, *11*, 600.
63. Huang, Q.; Jiang, F.; Wang, L.; Yang, C. Design of photobioreactors for mass cultivation of photosynthetic organisms. *Engineering* **2017**, *3*, 318–329. [CrossRef]
64. Xu, J.; Ma, B.; Su, J.; Huang, S.; Xu, X.; Zhou, X.; Huang, X.E.; Knight, R. Emerging trends for microbiome analysis: From single-cell functional imaging to microbiome Big Data. *Engineering* **2017**, *3*, 66–70. [CrossRef]
65. Oliveira, A.; Navia, R. Zero waste bio-refineries: The example of Brazilian pulp and paper sector. *Waste Manag. Res.* **2018**, *35*, 1193–1194. [CrossRef] [PubMed]

66. Guo, X.; Kang, K.; Shang, G.; Yu, X.; Qiu, L.; Sun, G. Influence of mesophilic and thermophilic conditions on the anaerobic digestion of food waste: Focus on the microbial activity and removal of long chain fatty acids. *Waste Manag. Res.* **2018**, *36*, 1106–1112. [CrossRef]
67. Ahorsu, R.; Medina, F.; Constantí, M. Significance and challenges of biomass as a suitable feedstock for bioenergy and biochemical production: A Review. *Energies* **2018**, *11*, 3366. [CrossRef]
68. Ragossnig, A.; Tunesi, S. Succeeding in involving decision-makers in solid waste management planning by setting appropriate goals and providing a structured planning method. *Waste Manag. Res.* **2018**, *36*, 924–933. [CrossRef]
69. Chen, F.; Luo, Z.; Yang, Y.; Liu, G.J.; Ma, J. Enhancing municipal solid waste recycling through reorganizing waste pickers: A case study in Nanjing, China. *Waste Manag. Res.* **2018**, *36*, 767–778. [CrossRef]
70. Müller, F.P.C.; Maack, G.-C.; Buescher, W. Effects of biogas substrate recirculation on methane yield and efficiency of a liquid-manure-based biogas plant. *Energies* **2017**, *10*, 325. [CrossRef]
71. Mortreuil, P.; Baggio, S.; Lagnet, C.; Schraauwers, B.; Monlau, F. Fast prediction of organic wastes methane potential by near infrared reflectance spectroscopy: A successful tool for farm-scale biogas plant monitoring. *Waste Manag. Res.* **2018**, *36*, 800–809. [CrossRef] [PubMed]
72. Llewellyn, D. Does global agriculture need another green revolution? *Engineering* **2018**, *4*, 449–451. [CrossRef]
73. Anjum, M.; Khalid, A.; Qadeer, S.; Miandad, R. Synergistic effect of co-digestion to enhance anaerobic degradation of catering waste and orange peel for biogas production. *Waste Managem. Res.* **2017**, *35*, 967–977. [CrossRef]
74. Roubík, H.; Mazancová, J. Identification of context-specific knowledge as a tool for biogas facilitators and their quality involvement—Using Vietnamese practice as an example. *Energies* **2019**, *12*, 1326. [CrossRef]
75. Hildebrandt, J.; Bezama, A. Cross-fertilisation of ideas for a more sustainable fertiliser market: The need to incubate business concepts for harnessing organic residues and fertilisers on biotechnological conversion platforms in a circular bioeconomy. *Waste Manag. Res.* **2018**, *36*, 1125–1126. [CrossRef]
76. Franco, R.T.; Coarita, H.; Bayard, R.; Buffière, P. An improved procedure to assess the organic biodegradability and the biomethane potential of organic wastes for anaerobic digestion. *Waste Manag. Res.* **2019**, *37*, 746–754. [CrossRef]
77. Kannavou, M.; Zampara, M.; Capros, P. Modelling the EU internal electricity market: The PRIMES-IEM Model. *Energies* **2019**, *12*, 2887. [CrossRef]
78. Christidis, P.; Focas, C. Factors affecting the uptake of hybrid and electric vehicles in the European Union. *Energies* **2019**, *12*, 3414. [CrossRef]
79. Statharas, S.; Moysoglou, Y.; Siskos, P.; Zazias, G.; Capros, P. Factors influencing electric vehicle penetration in the EU by 2030: A Model-Based Policy Assessment. *Energies* **2019**, *12*, 2739. [CrossRef]
80. Achinas, S.; Achinas, V. Biogas combustion: An introductory briefing. In *Biogas: Production, Applications and Global Developments*; Vico, A., Artemio, N., Eds.; Nova Science Publishers, Inc.: New York, NY, USA, 2017; pp. 179–193.
81. Gao, Y.; Kong, X.; Xing, T.; Sun, Y.; Zhang, Y.; Luo, X.; Sun, Y. Digestion performance and microbial metabolic mechanism in thermophilic and mesophilic anaerobic digesters exposed to elevated loadings of organic fraction of municipal solid waste. *Energies* **2018**, *11*, 952. [CrossRef]
82. Wagner, O.; Lackner, N.; Mutschlechner, M.; Prem, E.M.; Markt, R.; Illmer, P. Biological pretreatment strategies for second-generation lignocellulosic resources to enhance biogas production. *Energies* **2018**, *11*, 1797. [CrossRef] [PubMed]
83. Zielińska, K.; Fabiszewska, A.; Świątek, M.; Szymanowska-Powałowska, D. Evaluation of the ability to metabolize 1,2-propanediol by heterofermentative bacteria of the genus Lactobacillus. *Electron. J. Biotechnol.* **2017**, *26*, 60–63.
84. Nelson, M.J.; Nakhla, G.; Zhu, J. Fluidized-bed bioreactor applications for biological wastewater treatment: A review of research and developments. *Engineering* **2017**, *3*, 330–342. [CrossRef]
85. Okoro, O.V.; Sun, Z.; Birch, J. Prognostic assessment of the viability of hydrothermal liquefaction as a post-resource recovery step after enhanced biomethane generation using co-digestion technologies. *Appl. Sci.* **2018**, *8*, 2290. [CrossRef]
86. Achinas, S.; Euverink, G.J.W. Effect of combined inoculation on biogas production from hardly degradable material. *Energies* **2019**, *12*, 217. [CrossRef]

87. Boonpiyo, S.; Sittijunda, S.; Reungsang, A. Co-digestion of napier grass with food waste and napier silage with food waste for methane production. *Energies* **2018**, *11*, 3200. [CrossRef]
88. Wu, Y.; Li, T.; Cao, Q.; Li, X.; Zhang, Y.; Tan, X. RecET recombination system driving chromosomal target gene replacement in Zymomonas mobilis. *Electron. J. Biotechnol.* **2017**, *30*, 118–124. [CrossRef]
89. Chiumenti, A.; Boscaro, D.; da Borso, F.; Sartori, L.; Pezzuolo, A. Biogas from fresh spring and summer grass: Effect of the harvesting period. *Energies* **2018**, *11*, 1466. [CrossRef]
90. Sadala, S.; Dutta, S.; Raghava, R.; Jyothsna, T.S.; Chakradhar, B.; Ghosh, S.K. Resource recovery as alternative fuel and raw material from hazardous waste. *Waste Manag. Res.* **2019**, *37*, 1063–1076. [CrossRef]
91. Mamimin, C.; Prasertsan, P.; Kongjan, P.; O-Thong, S. Effects of volatile fatty acids in biohydrogen effluent on biohythane production from palm oil mill effluent under thermophilic condition. *Electron. J. Biotechnol.* **2017**, *29*, 78–85. [CrossRef]
92. Świątek, M.; Soo, C.S.; Yap, W.S.; Hon, W.M.; Ramli, N.; Kalsom Md Shah, U.; Phang, L.Y. Improvement of hydrogen yield of ethanol-producing Escherichia coli recombinants in acidic conditions. *Electron. J. Biotechnol.* **2017**, *26*, 27–32.
93. Marcos, A.C.; Al-Kassir, A.; Cuadros, F.; Yusaf, T. Treatment of slaughterhouse waste water mixed with Serum from Lacteal Industry of Extremadura in Spain to Produce Clean Energy. *Energies* **2017**, *10*, 765. [CrossRef]
94. Morales-Polo, C.; Cledera-Castro, M.D.M.; Moratilla Soria, B.Y. Reviewing the anaerobic digestion of food waste: From waste generation and anaerobic process to its perspectives. *Appl. Sci.* **2018**, *8*, 1804. [CrossRef]
95. Mangoyana, R.B.; Smith, T.F.; Simpson, R. A systems approach to evaluating sustainability of biofuel systems. *Renew. Sustain. Energy Rev.* **2013**, *25*, 371–380. [CrossRef]
96. Eisentraut, A. Sustainable Production of Second-Generation Biofuels. In *Potential and Perspectives in Major Economies and Developing Countries*; Internation Energy Agency: Paris, France, 2010.
97. Lazaroiu, G.; Mihaescu, L.; Negreanu, G.; Pana, C.; Pisa, I.; Cernat, A.; Ciupageanu, D.-A. Experimental investigations of innovative biomass energy harnessing solutions. *Energies* **2018**, *11*, 3469. [CrossRef]
98. Černý, M.; Vítězová, M.; Vítěz, T.; Bartoš, M.; Kushkevych, I. Variation in the distribution of hydrogen producers from the Clostridiales order in biogas reactors depending on different input substrates. *Energies* **2018**, *11*, 3270. [CrossRef]
99. Chen, J.F. Green chemical engineering. *Engineering* **2017**, *3*, 283–284. [CrossRef]
100. Nelabhotla, A.B.T.; Dinamarca, C. Bioelectrochemical CO_2 reduction to methane: MES integration in biogas production processes. *Appl. Sci.* **2019**, *9*, 1056. [CrossRef]
101. Macedonio, F.; Drioli, E. Membrane engineering for green process engineering. *Engineering* **2017**, *3*, 290–298. [CrossRef]
102. Diaz-Chavez, R.A. Assessing biofuels: Aiming for sustainable development or complying with the market? *Energy Policy* **2011**, *39*, 5763–5769. [CrossRef]
103. Junginger, M.; Van Dam, J.; Zarrilli, S.; Mohamed, F.A.; Marchal, D.; Faaij, A. Opportunities and barriers for international bioenergy trade. *Energy Policy* **2011**, *39*, 2028–2042. [CrossRef]
104. Festel, G.; Würmseher, M.; Rammer, C. Scaling and Learning Effects of Biofuels Conversion Technologies. *Energy Technol.* **2004**, *2*, 612–617. [CrossRef]
105. De Souza Guimarães, C.; da Silva Maia, D.R.; Serra, E.G. Construction of biodigesters to optimize the production of biogas from anaerobic co-digestion of food waste and sewage. *Energies* **2018**, *11*, 870. [CrossRef]
106. Nitsche, M.; Hensgen, F.; Wachendorf, M. Using grass cuttings from sports fields for anaerobic digestion and combustion. *Energies* **2017**, *10*, 388. [CrossRef]
107. Li, W.-W.; Yu, H.-Q. Advances in energy-producing anaerobic biotechnologies for municipal wastewater treatment. *Engineering* **2016**, *2*, 438–446. [CrossRef]
108. Solarte-Toro, J.C.; Chacón-Pérez, Y.; Cardona-Alzate, C.A. Evaluation of biogas and syngas as energy vectors for heat and power generation using lignocellulosic biomass as raw material. *Electron. J. Biotechnol.* **2018**, *33*, 52–62. [CrossRef]
109. Wang, J.; Wang, H.; Fan, Y. Techno-economic challenges of fuel cell commercialization. *Engineering* **2018**, *4*, 352–360. [CrossRef]
110. Seneesrisakul, K.; Sutabutr, T.; Chavadej, S. The effect of temperature on the methanogenic activity in relation to micronutrient availability. *Energies* **2018**, *11*, 1057. [CrossRef]

111. Baek, G.; Kim, J.; Kim, J.; Lee, C. Role and potential of direct interspecies electron transfer in anaerobic digestion. *Energies* **2018**, *11*, 107. [CrossRef]
112. Franchi, O.; Rosenkranz, F.; Chamy, R. Key microbial populations involved in anaerobic degradation of phenol and p-cresol using different inocula. *Electron. J. Biotechnol.* **2018**, *35*, 33–38. [CrossRef]
113. Musa, M.A.; Idrus, S.; Che Man, H.; Nik Daud, N.N. Wastewater treatment and biogas recovery using Anaerobic Membrane Bioreactors (AnMBRs): Strategies and Achievements. *Energies* **2018**, *11*, 1675. [CrossRef]
114. Lemões, J.S.; e Silva, C.F.L.; Avila, S.P.F.; Montero, C.R.S.; e Silva, S.D.D.A.; Samios, D.; do Carmo Ruaro Peralba, M. Chemical pretreatment of Arundo donax L. for second-generation ethanol production. *Electron. J. Biotechnol.* **2018**, *31*, 67–74. [CrossRef]
115. Duan, N.; Ran, X.; Li, R.; Kougias, P.G.; Zhang, Y.; Lin, C.; Liu, H. Performance evaluation of mesophilic anaerobic digestion of chicken manure with algal digestate. *Energies* **2018**, *11*, 1829. [CrossRef]
116. UNEP. *The Bioenergy and Water Nexus*; Oeko-Institut and IEA Bioenergy Task 43; UNEP: Nairobi, Kenya, 2011.
117. Koizumi, T. Biofuels and food security. *Renew. Sustain. Energy Rev.* **2015**, *52*, 829–841. [CrossRef]
118. European Commission. *Renewable Energy in Europe—2018*; European Environment Agency: Copenhagen, Denmark, 2018.
119. Cherubini, F.; Bird, N.D.; Cowie, A.; Jungmeier, G.; Schlamadinger, B.; Woess-Gallasch, S. Energy- and greenhouse gas-based LCA of biofuel and bioenergy systems: Key issues, ranges and recommendations. *Resour. Conserv. Recycl.* **2009**, *53*, 434–447. [CrossRef]
120. Cherubini, F.; Strømman, A.H. Life cycle assessment of bioenergy systems: State of the art and future challenges. *Bioresour. Technol.* **2011**, *102*, 437–451. [CrossRef]
121. Havlik, P.; Schneider, U.A.; Schmid, E.; Böttcher, H.; Fritz, S.; Skalský, R.; Aoki, K.; De Cara, S.; Kindermann, G.; Kraxner, F.; et al. Global land-use implications of first and second generation biofuel targets. *Energy Policy* **2011**, *39*, 5690–5702. [CrossRef]
122. Overmars, K.P.; Stehfest, E.; Ros, J.P.; Prins, A.G. Indirect land use change emissions related to EU biofuel consumption: An analysis based on historical data. *Environ. Sci. Policy* **2011**, *14*, 248–257. [CrossRef]
123. Hellmann, F.; Verburg, P.H. Impact assessment of the European biofuel directive on land use and biodiversity. *J. Environ. Manag.* **2010**, *91*, 1389–1396. [CrossRef]
124. International Renewable Energy Agency. *Innovation Outlook: Advanced Liquid Biofuels*; International Renewable Energy Agency: Abu Dhabi, UAE, 2016.
125. Festel, G.; Würmseher, M.; Rammer, C.; Boles, E.; Bellof, M. Modelling production cost scenarios for biofuels and fossil fuels in Europe. *J. Clean. Prod.* **2014**, *66*, 242–253. [CrossRef]
126. FAO. *FAOSTAT Database*; Food and Agriculture Organization of the United Nations: Rome, Italy, 2017.
127. Millinger, M.; Thrän, D. Biomass price developments inhibit biofuel investments and research in Germany; The crucial future role of high yields. *J. Clean. Prod.* **2018**, *172*, 1654–1663. [CrossRef]
128. Strengers, B.; Overmars, K.; Kram, T.; Ros, J. *Greenhouse Gas Impact of Bioenergy Pathways*; PBL Netherlands Environmental Assessment Agency: The Hague, The Netherlands, 2016.
129. European Commission. *The Land Use Change Impact of Biofuels Consumed in the EU Quantification of Area and Greenhouse Gas Impacts*; European Commission: Utrecht, The Netherlands, 2015.
130. European Commission. *Positions, Recommendations and Key Messages from the Industry*; European Commission: Luxembourg, 2017.
131. Bitnere, K.; Searle, S. *Effective Policy Design for Promoting Investment in Advanced Alternative Fuels*; International Council on Clean Transportation: Washington, DC, USA, 2017.
132. European Biogas Association. *Biomethane in Transport*; European Biogas Association: Brussels, Belgium, 2016.
133. JRC; EUCAR; CONCAWE. BioGrace Excel Tool version 4d. In *Harmonised Calculations of Biofuel Greenhouse Gas Emissions in Europe*; BioGrace: Brussels, Belgium, 2015.
134. Hamelinck, C.; Hoogwijk, M. *Future Scenarios of First and Second Generation Biofuels*; Ecofys MNP: Utrecht, The Netherlands, 2007.
135. Tonini, D.; Hamelin, L.; Alvarado-Morales, M.; Astrup, T.F. GHG emission factors for bioelectricity, biomethane, and bioethanol quantified for 24 biomass substrates with consequential life-cycle assessment. *Bioresour. Technol.* **2016**, *208*, 123–133. [CrossRef] [PubMed]
136. European Commission. *Reporting Requirements on Biofuels and Bioliquids Stemming from the Directive (EU) 2015/1513*; European Commission: Brussels, Belgium, 2017.

137. European Commission. *Estimates of Indirect Land Use Change from Biofuels Based on Historical Data*; JRC Science and Policy Reports: Ispra, Italy, 2015.
138. Rulli, M.C.; Bellomi, D.; Cazzoli, A.; De Carolis, G.; D'Odorico, P. The Water-Land-Food Nexus of First-Generation Biofuels. *Sci. Rep.* **2016**, *6*, 22521. [CrossRef] [PubMed]
139. UK Department for Transport. *Renewable Transport Fuel Obligation Statistics: Period 10 (2017/18)*; UK Department for Transport: London, UK, 2019.
140. Mathioudakis, V.; Gerbens-Leenes, P.W.; Van der Meer, T.H.; Hoekstra, A.Y. The water footprint of second-generation bioenergy: A comparison of biomass feedstocks and conversion techniques. *J. Clean. Prod.* **2017**, *148*, 571–582. [CrossRef]
141. Immerzeel, D.J.; Verweij, P.A.; Van der Hilst, F.; Faaij, A.P.C. Biodiversity impacts of bioenergy crop production; a state-of-the-art review. *Glob. Chang. Biol. Bioenergy* **2014**, *6*, 183–209. [CrossRef]
142. Kline, K.L.; Martinelli, F.S.; Mayer, A.L.; Medeiros, R.; Oliveira, C.O.F.; Sparovek, G.; Walter, A.; Venier, L.A. Bioenergy and Biodiversity: Key Lessons from the Pan American Region. *Environ. Manag.* **2015**, *56*, 1377–1396. [CrossRef]
143. Selbmann, K.; Pforte, L. Evaluation of Ecological Criteria of Biofuel Certification in Germany. *Sustainability* **2016**, *8*, 936. [CrossRef]
144. Su, Y.; Zhang, P.; Su, Y. An overview of biofuels policies and industrialization in the major biofuel producing countries. *Renew. Sustain. Energy Rev.* **2015**, *50*, 991–1003. [CrossRef]
145. Ministry of Energy (PL). *Interim Report on Progress in the Promotion and Use of Energy from Renewable Sources in Poland in 2015–2016*; Ministry of Energy Poland: Warsaw, Poland, 2018.
146. The Netherlands Enterprise Agency. Progress report. In *Energy from Renewable Sources in The Netherlands 2015–2016*; The Netherlands Enterprise Agency: Utrecht, The Netherlands, 2017.
147. McGrath, M. Italy Pushes Ahead with 'Next Generation' Biofuels from Waste. 2014. Available online: https://www.bbc.com/news/science-environment-29618889 (accessed on 27 May 2019).
148. Ministry of Energy (RO). Romania's Progress Report on Promoting and Using Energy from Renewable Sources, in Compliance with Article 22 of Directive 2009/28/EC. Available online: https://www.eurobserv-er.org/pdf/res-policy/EurObservER-RES-Policy-Report-Article-22-Romania-report-EN.pdf (accessed on 27 August 2019).
149. Ecofys. *Technical Assistance in Realisation of the 2018 Report on Biofuels Sustainability: Biofuels, Biomass & Biogas Used for Renewable Energy Generation*; Ecofys: Utrecht, The Netherlands, 2019.
150. Krukowska, E. EU Sets Limits on Palm Oil in Biofuels as Trade War Looms. 2019. Available online: https://www.bloomberg.com/news/articles/2019-03-13/eu-trying-to-go-green-and-avoid-trade-war-with-palm-oil-giants (accessed on 27 May 2019).
151. Beckman, J. *Biofuel Use in International Markets: The Importance of Trade*; U.S. Department of Agriculture, Economic Research Service: Washington, DC, USA, 2015.

sustainability

MDPI

Article

The Role of Participation in the Responsible Innovation Framework for Biofuels Projects: Can It Be Assessed?

Andreia Marques Postal [1,2,*], Gabriela Benatti [1,2], Mar Palmeros Parada [2], Lotte Asveld [2], Patrícia Osseweijer [2] and José Maria F. J. Da Silveira [1]

1 Department of Agricultural and Environmental Studies, Institute of Economics, University of Campinas, Pitágoras 353, Campinas CEP 13083-857, Brazil; g.solidariodesouzabenatti@tudelft.nl (G.B.); jmsilv@unicamp.br (J.M.F.J.D.S.)

2 Department of Biotechnology, Faculty of Applied Sciences, Delft University of Technology, Van der Maasweg 9, 2629 HZ Delft, The Netherlands; m.d.m.palmerosparada@tudelft.nl (M.P.P.); l.asveld@tudelft.nl (L.A.); p.osseweijer@tudelft.nl (P.O.)

* Correspondence: andreiamarquesp@gmail.com or a.camargomarquespostal@tudelft.nl; Tel.: +55-19-9813-10-303

Received: 30 October 2020; Accepted: 10 December 2020; Published: 17 December 2020

Abstract: The growth in biofuels' investment brings with it concerns about the social and environmental impacts of the sector. Several tools and frameworks have been used to address these concerns, including the Responsible Research and Innovation (RRI) framework. This paper analyzes whether this framework can be applied in contexts where local culture and values shape differently the freedom of speech and engagement, such as in developing countries in which biofuel innovation projects are often implemented. A literature review focused on eight case studies of other authors was used to explore the role of "participation" as a structural element of the RRI framework and the impact of its absence where effective participation in the innovation development process is not possible. In conclusion, we highlight how this inspirational normative framework, designed to influence innovation, is misused to judge its impacts. More than that, the conclusions of such misused applications reflect more the difficulties involved in applying guidelines than the responsible character of the innovation, whose impacts are usually defined upfront materially and measurably.

Keywords: responsible research and innovation; biofuels; effective participation; inclusion; institutions

1. Introduction

There is a growing demand for alternative, renewable energies. One of them is biofuels, which are well-suited to the transport sector due to their liquid form and the diversity of raw materials and potential local production sources. However, along with the growth in this sector's investments, concerns have emerged regarding social and environmental impacts, such as land-use changes, food security, and biodiversity loss.

Different frameworks and assessment tools have emerged to analyze these social and environmental impacts, among them, the Responsible Research and Innovation Framework (RRI). The RRI framework was first employed in Europe after the economic crisis of 2008 and introduced the consideration of values and concerns related to innovation. As an inspirational normative framework aimed to influence researchers and innovators through its four pillars, the framework played a relevant role in the development of research and innovation in Europe since it was adopted by important research institutions in The Netherland and UK and is a transversal guiding principle of the "Horizon 2020 Research and Innovation Program" from the European Union [1–6].

However, biofuel projects are implemented in Europe's technological innovation centers and the US and countries far from the developed countries where the framework was created. As different cultures and values have different ways of interacting and building solutions, we must consider whether the use of the RRI framework truly reflects the responsible nature of innovation itself or whether it reflects the difficulties involved in applying guidelines that were, perhaps unconsciously, embedded with western values. Section 2 presents how the analysis through a literature review focused on case studies (from the biofuel and other sectors). The research about the framework's structural elements was used to complement the analysis and reach more accurate conclusions about the responsible character of innovations.

Section 3 describes the growing importance of biofuels in the transition towards a biobased society and some of this sector's characteristics. Section 4 is dedicated to exploring the origin and aims of the RRI framework to analyze innovation's responsible character. Aspects such as the participatory process and stakeholder engagement are at the core of the framework's recommended activities. However, the inclusiveness dimension encapsulates them and, because of that, demand more attention to the effectiveness of all other dimensions. These reflections highlight the processual meaning of the term "inclusion", as well as the meaning of the term "responsible" and the impact of these characteristics on the application of RRI in contexts where different values and behaviours shape matters of participation and engagement.

In Section 5, we introduce the "institutional approach" as a possible complementary analysis to understand these different contextual elements which shape the participatory process and the application of RRI. Institutions have a dual role: as "action arenas" or deliberative forums where stakeholder discussions take place and acquire deliberative functions and coercion; at the same time, institutions are key elements in understanding the local context, culture, values, and behaviors. Section 6 discusses what happens when such institutions are not in place, illustrating several case studies carried out by other authors. These cases are analyzed to understand RRI applicability's limits as an assessment and the institutional analysis's complementary role.

Finally, we describe the research's main conclusions about using RRI in assessments to describe the "responsible" character of innovation and complementary ways to improve the analysis.

2. Methodology

We conducted a literature review using the Web of Science and Science Direct. The search terms on topics included "responsible research and innovation," "responsible research and innovation AND biofuel", "responsible research and innovation AND bioeconomy", "responsible research and innovation AND developing countries", "responsible research and innovation AND global South", "responsible research and innovation AND southern countries", and "responsible research and innovation AND South America". An overview of the search procedures is presented in Figure 1.

Details about the selected studies are in Table 1. The articles are presented in the table by alphabetic order of authorship.

Table 1. Articles selected.

Authors	Title	Year of Publication	Studied Location	Innovation Focus
Balkema, A.; Pols, A.	Biofuels: Sustainable Innovations or Gold Rush? Identifying Responsibilities form Biofuels Innovations	2015	Tanzania	Jatropha biofuels
Chaturvedi et al.	Agriculture Technology Choices and the (RRI) Framework: Emerging Experiences from China and India	2016	India and China	Transgenic cotton
De Campos et al.	Responsible Innovation and political accountability: genetically modified mosquitoes in Brazil	2017	Brazil	Genetically modified Aedes Aegypt
De Hoop et al.	Limits to responsible Innovation	2016	India	Jatropha biofuels
Hartley et al..	A retrospective analysis of responsible Innovation for low-technology Innovation in the Global South	2019	Europe and Africa	Low technologies
Macnaghten et al.	Responsible Innovation across borders: tensions, paradoxes and possibilities	2014	Cross-cultural	RRI from a cross-cultural perspective
Vasen	Responsible Innovation in Developing Countries: An Enlarged Agenda	2017	Developing countries	RRI in Developing countries
Voeten, J.J.; Naudé, W.A	Regulating the negative externalities of enterprise cluster innovations: lessons from Vietnam	2014	Vietnam	Small informal enterprises

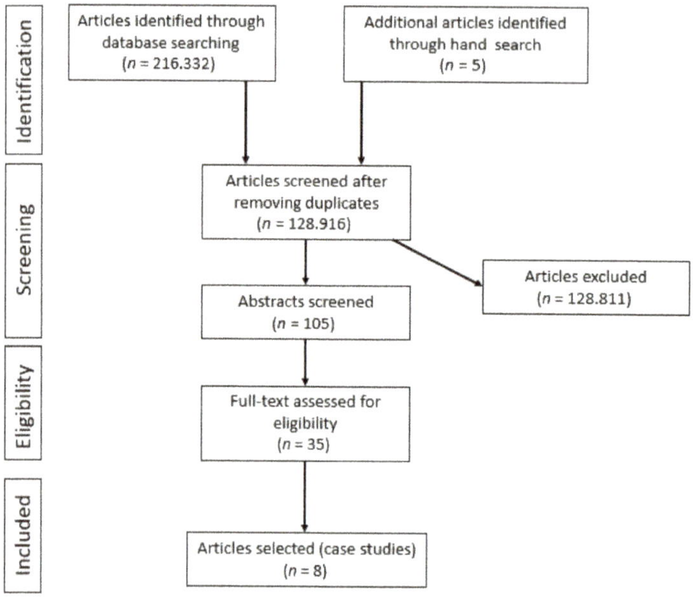

Figure 1. Search procedures.

3. Biofuels as Cleaner Energy and Global Commodity

The worldwide concern about fossil fuels, both in terms of their environmental impact and energy security, has resulted in searching for alternative energy sources to replace or supplement countries' energy matrix. Such alternative energies include hydropower, wind, solar, geothermal, and biofuels, as presented in Figure 2.

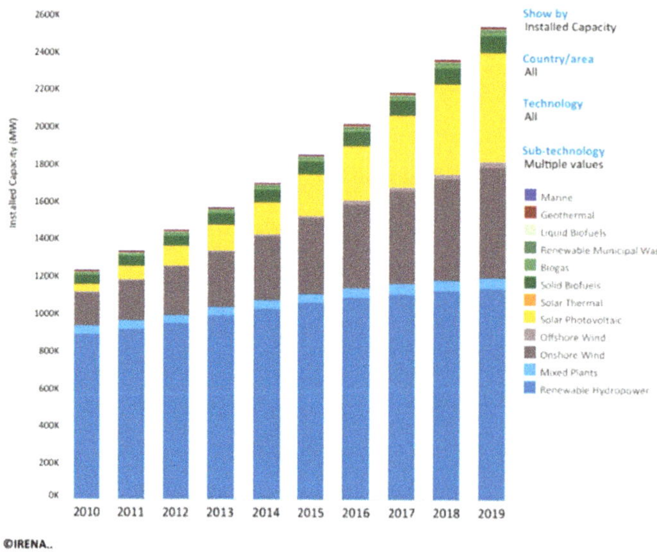

Figure 2. Global installed capacity of renewable energies [7].

Biofuels have been developed for some time now as a renewable, drop-in alternative for fossil fuels. They are well-suited for use in the transport sector. Their liquid form favours transportation to the final consumer and can be easily incorporated into the existing distribution infrastructure, for example, through blending [8]. This makes biofuels an essential energy source in the transition from fossil fuels to clean energy. However, the global renewable energy debate is now shared with other alternatives, especially wind and solar PV, whose investments have grown significantly [9,10]. Support for renewable energies and energy efficiency in road transportation largely occurs by incentivizing biofuels, electric cars, and fuel efficiency. Another benefit of biofuels is that they can be obtained from various raw materials produced in a range of locations, which may improve countries' autonomy and energy security and reduce their dependence on fossil fuels.

Many countries now have blending targets in place, which has led to an increase in the demand for liquid biofuels, especially in the United States, the European Union (EU), and Japan, and reflected a rise in the trade and import of biofuels, such as Brazilian ethanol, Jatropha, palm oil, and other sources of vegetable oil and biodiesel, from the United States, Latin America, and Southeast Asia. The production of biofuels has created new markets, and energy trade flows worldwide [11]. Based on the international liquidity of capital during the 2000s and the increased demand for biofuels worldwide, several investments were made in different countries, in other crops and contexts.

3.1. Concerns about Biofuel Expansion

However, the growing demand and investment in biofuel production implied new environmental and social concerns. Questions about the impact of biofuels on land-use changes [12–14], food security [15–18], deforestation and greenhouse gas emissions [19–22], have formed the core of the biofuel expansion debate [23–25]. The literature recognizes different types of impact depending on the source of the biofuels (1st generation, 2nd generation or biodiesel). However, the usefulness of an assessment approach to verify the extension of those impacts keep useful regardless the type of the biofuels in question.

On the other hand, positive aspects have been added to the food/energy/forest "trilemma" debate [26–28] to analyze the trade-offs involved. Examples include job creation and improved working conditions in the value chains [29–32], rural development [33–35] and integration with smallholder farming practices [36]. Therefore, if biofuel production is to attain sustainable decarbonization and other renewable energy targets, the societal context must prevent or minimize negative outcomes and increase benefits.

3.2. How to Address These Concerns

In Brazil, where developments in the bioenergy sector are centered on advanced biofuels (first-and-second-generation ethanol), interest has been shown from relevant sectors to address these concerns. Emerging bioenergy sectors include, for example, second-generation ethanol, new drop-in applications such as jet fuels, and the diversification of biodiesel sources and technologies for lignin extraction. The processes involve mature technologies and intermediate and emergent sectors [37,38]. This interest is illustrated by the growing partnerships between biofuel developers and users, and societal actors such as NGOs, as well as their participation in voluntary sustainability certifications such as the Better Sugarcane Initiative, now Bonsucro (https://www.bonsucro.com/), Roundtable on Sustainable Biofuels (https://rsb.org/) (RSB), among others [31,39–41].

One way to address societal concerns is to influence the researchers and entrepreneurs when the innovation is at the initial stages of development. In Europe and the United States, Responsible Research and Innovation (RRI) guidelines were drawn up to increase all interested stakeholders' participation to promote the alignment of social needs and values with the technology being developed [42,43]. Besides its use as an analytical framework, the RRI approach can address emerging societal and environmental concerns in ongoing developments in the biofuel sector or any other emerging technology field. The implementation of RRI is not only in the interest of the public and the RRI community but also in the biofuel sector's interest in addressing concerns.

However, since many biofuel projects are implemented in developing countries in the Global South [4,44,45], it is not clear whether the RRI approach can be used in contexts other than Europe for this purpose. In the following section, we describe the main concepts of the RRI framework and the reasons for these doubts.

4. The RRI Framework and the Role of "Participation" in Innovation Projects

The RRI approach fundamentally looks to generate ethical practice and awareness by influencing emerging technologies in innovation research and ventures. The strategy to do so is to inspire the innovation process through the effective participation of a range of stakeholders that guarantee the consideration of sustainable aspects in socially desirable projects and, in doing so, respond to the main problems faced by humanity [1,42,43,46]. Its design emphasizes that innovation is a collective action involving several agents and institutions engaged at all stages of the process, from agenda-setting to design, implementation, and evaluation. Agents include researchers, policymakers, educators, entrepreneurs, and civil society organizations, encompassing the people directly impacted by the innovation. These ideas were promoted by the EU and incorporated, for example, into the Dutch Research Council (NWO) guidelines, and as a result, spread throughout the academic arena in Europe and the US [47]. It is important to note that the societal context that shaped this framework was marked by essential debates in society on ethics, the role of science and innovation, sustainability, climate change, nanotechnology, and increasing demand for participation in the decision-making process in democratic societies [48].

However, the framework presents some limitations [1,49–52]. Its normative character is considered one reason why it doesn't have a central role in Horizon Europe. However, its principles and aims remain relevant in the context of this new phase of the European research agenda and academic debates. In this section, we aim to analyze how "effective participation" is a core attribute for its four pillars' efficiency and, therefore, a reason why using the scheme as an assessment tool is inappropriate. This is because local culture and historical contexts shape the ways of interaction in each country or region differently and, therefore, produce variable results in the process of participation and applicability of RRI pillars, which may have no connection to the positive impact of the innovation itself.

4.1. The Main Pillars of RRI

The framework was built based on four main pillars that continuously reinforce each other. The proposed dimensions have been broadly accepted as requirements for responsible research and innovation through normative guidelines, although many authors have made amendments and additions. The framework is meant to be used as a proactive and inspirational guideline to meet societal expectations. Therefore, its use in a retrospective way should be avoided or, at least, used with care to avoid conclusions and judgments over innovation impacts while assessing just the innovation development process. The same holds for actors and activities that were not planned to be aligned with such a framework built afterward. We also question its use as an assessment tool due to the lack of objectivity and defined comparable parameters.

According to Stilgoe et al. [43], the four dimensions are:

- Anticipation: to consider what we know and what is probable in terms of technology use. Anticipation aims to foresee the potential risks of the new technology and require a "well-timed" approach to be meaningful and constructive. Another essential requirement is "transparency" since the shared knowledge and information are crucial to enabling participants to assess the risks and future impact of innovations [1].
- Reflexivity: to recognize the moral views, interests, and roles of all relevant stakeholders and critically examine the assumptions behind them. Stilgoe et al. [43] also highlight that reflexivity should be present not only at the laboratory level but mainly at the "institutional" level (research funders, regulators, and other institutions). These organizations should help build the reflexive capacity

among engaged stakeholders and reflect on their value system, shaping the innovation process. To do so, the independent thinking of the participants is crucial.

- Responsiveness: the capacity to respond to the questions and insights raised during the innovation process. In this sense, the "actors" involved should have enough power, resources, capabilities, and skills to do so. According to Stilgoe et al. [43], it implies the notion of innovations being shaped so that they are "as responsive as possible" to the adjustments that occur throughout their development while recognizing the insufficiency of knowledge and control. However, the authors point out the need for diversity (of stakeholders) to have a responsive innovation system.
- Inclusion: refers to how relevant stakeholders are engaged in the innovation process. To achieve qualified outcomes of the inclusiveness process, Stilgoe et al. [43] refer to three criteria as proposed by Callon [53]: intensity, or how actors are invited to participate promptly; openness, in other words, how the group is composed and how diverse it is; and the quality of the discussion which depends on the participants' critical thinking skills.

One crucial challenge highlighted by the author regards the power imbalance among the participants, which can jeopardize discussion outcomes. Additionally, issues such as confidentiality, information asymmetry, and power imbalance are frequently observed during the use of the approach. All these elements highlight the importance of effective participation to achieve "substantive inclusion," which can achieve the framework's aimed outcomes [49].

4.2. RRI Demands an Effective Participatory Process

Through its architecture and dimensions, the RRI framework emphasizes inclusion as a process of participation and engagement, and this is based on a belief that, if this processual inclusion is well-executed (effective participation), we will have a substantive inclusion [49], the stakeholders will together be able to build the best solution (or material results), taking into account society's needs and the interests of all groups. This characterizes RRI as a processual framework where the stakeholder's inclusion and participation are key to achieve the desired societal outcomes brought to the table by these participants [50]. Participation is, therefore, a key element to the successful execution of all RRI dimensions. Still, it requires the right set of stakeholders to produce results aligned with the RRI framework effectively. The right set of agents, who would be chosen or invited by trustable organizations, would have characteristics such as representing a diverse range of stakeholders, consisting of independent skilled people with a high level of critical thinking who can discuss and avoid manipulation. Only the right set of stakeholders in a process coordinated by a trustable organization would deliver the anticipation, reflexiveness, and responsiveness dimensions required by the framework.

Therefore, it is reasonable to deduce that the term "responsible", as pursued by the RRI theoretical framework, is directly related to organizing a participatory process (by inviting relevant audiences who would perform the four pillar activities) effectively. This "effective participation" could ensure the four pillar activities' satisfactory performance while indirectly achieving the innovation process's material outcomes. What are the requirements for "effective participation" in the implementation of a biofuels project?

We understand that two requirements are essential to guarantee an effective participatory process to achieve the final goal of the RRI approach: (a) an adequate set of stakeholders and (b) adequate institutions for coordinating the process and working as an "action arena with deliberative power to achieve"responsible innovation impacts.

The former, referring to the set of stakeholders, is relevant since the outcomes of each one of the four pillars are not given upfront but are defined by the participation and positioning of the group of invitees representing their interests. It can also be extracted from an in-depth analysis of the four pillars' implementation [43]. Each one of the posts requires specific attention to the definition of representativeness to attain the best results. Thus, the anticipation pillar requires a well-informed and diverse range of participants who can challenge and question potential risks. The reflexiveness pillar requires stakeholders' independence, the ability to rethink moral values, enriched with different stakeholders' perspectives. The responsiveness pillar requires openness, transparency, and response capabilities, which can only be attained through the participants' assertive group invited to the process. Finally, the inclusiveness itself requires different views from a range of interests and fields (scientists, governments, laymen, community leaders, and external experts) with different perspectives and a good representation of society.

Therefore, effective participation is a crucial element to discuss the future perspectives of emerging biofuel technologies, which are inherently uncertain and depend on the set of representatives' best configuration. Thus, it is necessary to consider the requirements these actors must fulfill to reach the process's best outcomes.

In fact, "effective participation" assumes that some values are present or given in any context or society [48]. However, the forms and patterns of interaction among stakeholders seem to differ in each country or region [51]. It is crucial to understanding how the local context, culture, and beliefs interfere with and shape the discussions on achieving the desired responsible innovation. This is particularly important when discussing research and innovation, which are developed in one place and implemented in another, as is usually the case of biofuels and other technological innovations [48,52,54,55].

Effective participation is related to the definition of groups and names and a second important element: the institutions where the discussions take place. This element from institution analysis theory [56] is usually responsible for answering organizational questions: when does the discussion occur? Is the frequency of meetings sufficient? Are the meetings multidimensional or just one-way communication events? Where do these meetings take place? Are all stakeholders able to attend? How do people interact with each other? How much freedom do they have to speak freely? How do gender and religion affect the ability to speak out and develop the right solution for everyone?

Answering these questions is out of this study's scope; nevertheless, they are mentioned here as they reveal the importance of understanding local practices. Some of these details are efficiently managed; others, however, are more difficult to deal with since they have hidden rules, behaviours, and beliefs that are difficult for an external stakeholder or an assessment body to understand and create global metrics to measure it, especially when discussing transcontinental projects where long distances impact closer coordination and recognition of context limitations [57]. In all cases, institutions are critical to effective participation and responsible innovation analysis. Because of this, their importance is further explored in the following section.

5. Institutions as Key Elements of the Innovation Process Analysis

> *Social and economic structures give rise to different configurations of actors in different countries at different times, who exercise influence not only on the making of policy but also on the making of institutions* [58].

5.1. Institutions as a Proxy for the Local Context

Barney et al. [59] define an institution as an organized pattern of constitutive collective behaviour and its evolution. In this sense, institutions are made up of values, norms, beliefs, meanings, symbols, customs, and socially learned and shared standards, which define the range of expected and accepted behaviour in a particular context. Aligned with this, Taebi et al. [52] explain that technological development occurs in specific institutional contexts involving individual stakeholder dynamics. These contexts, constituted by formal and informal institutions, reflect values that have significant ramifications for the governance structure, the distribution of power between stakeholders, transparency processes, gender equality, and the definition of priorities.

Echebarría et al [58] explains that institutions are path-dependent since the historical facts influence institutional evolution and its characterization. The institutional path and the social development path are not easily comparable or compatible with each other historical, economic, social, political, and cultural aspects influence the constitution of an institutional network, establishing specific and local forms of the development pattern [60]. All these factors set the limits within which the actors involved in these processes can act. Specifically, with regard to impact assessment, the mentioned factors can also limit the supervisory role of the responsible institutions. There are difficulties in carrying out impact analysis, financial and methodological challenges stand out. Once these studies are in place, even if measurable, there are also limitations in translating the results into norms and laws, and incorporating them into policies and the policy agenda.

In their place, important debates or frameworks can create resistance from stakeholders, as local organizations or society. The main reason is that, in general, these models are impregnated with values and realities that are very different from the local ones. In this sense, the network with other organizations and institutions is important to maintain objectives and inspire adaptations. Then, local multilateral organizations can act on adapting more general parameters to local realities at the more concrete level, we can mention agencies responsible for the regulation of biofuels, social policies and rules, even civil society represented by NGOs. The institutions, treated on a more general level of action arenas, manifest themselves in a dispute in the general plan of laws, rules, values and behaviours. This arena starts from the parliament, responsible for formulating rules and laws. In democracies, the members of the parliaments are elected, which theoretically, gives them independency to challenge agencies, institutions, organizations, and even the government [58].

Ostrom [61] also emphasizes the importance of local context, stating that exogenous variables such as (a) the biophysical and material conditions; (b) community attributes, including values, behaviours, and traditions; and (c) rules—formal and informal—that affect the action arenas (or institutions) where a group of participants (or stakeholders) interact with a specific purpose (or action situation). In this approach, "action arenas" shape and characteristics depend on the local contexts' exogenous variables.

Therefore, we argue that institutions are a proxy for analysing and incorporating local context specificities and that RRI should take these aspects into account. After all, each context manifests a specific reality that can impact the implementation of the dimensions and objectives of RRI. Incorporating these aspects would enable researchers and policymakers to shape and adapt the framework elements to the specific necessities regarding responsibility in the research and innovation processes.

Understanding the institutional contexts is important since limitations concerning the implementation of RRI may be more related to broader and deeply rooted socio-political factors and institutional contexts than to specific initiatives. Regarding this, Wickson and Carew [62] classify the "clear and explicit identification of institutional and contextual limitations and a structured effort to acknowledge and improve upon these conditions" as an exemplary criterion of reflexiveness and responsiveness.

5.2. Institutions as "Action Arenas" for Innovation Process Discussions

Another critical factor in incorporating an analysis of institutions into the RRI approach is that they represent the arena (or deliberative forum). The controversies surrounding and evolution of the innovation process occur. In Ostrom's words, an "action arena" is the "social space where individuals interact, exchange goods and services, solve problems, dominate one another, or fight" [56]. In sum, action arenas are collective spaces where the participation process takes place and, because of this, the characteristic of institutions is fundamental.

The question arises: which factors differentiate contexts with weak institutions from those with strong institutions? How is this linked to a potential RRI application? Levitsky and Murilo [63] highlight some features that support the differentiation of these institutional contexts: (1) socioeconomic inequality, (2) institutional borrowing, and (3) hasty institutional design.

Socioeconomic inequality is a factor that contributes to distancing society from important debates, particularly in contexts where there are low levels of education and access to information. There is a reduction in society's confidence in the institutions. Institutions are often imported from abroad without reflecting on the alignment with pre-existing domestic power structures or norms, leading to discrepancies. The rapid institutional design, in turn, affects institutional quality and durability, as there is inadequate time to anticipate impacts, include stakeholders, or further discuss issues.

Based on the argumentation described earlier, nations with a strong institutional configuration can conform to the RRI principles more easily than countries with a weak structure. The difficulty lies mostly in the concentration and imbalance of power, the lack of collective prioritization, aspects that impair representation (such as educational level), and political instabilities, to give some examples [63,64]. There may be no commitment to being "responsible" in these local contexts, one of the RRI framework aims.

To summarize, we have so far argued that (1) institutions are a proxy of local context and because of this can be used to improve strategies to implement RRI approach, and (2) institutions work as action arena or deliberative forum and are fundamental to coordinate the participatory innovation process. In the following section, we aim to illustrate these two points by presenting selected case studies from outside Europe (based on literature) that used RRI as an analysis tool to show how they failed to be objective and provide relevant comparisons. Several studies indeed illustrate the relevancy of the impact of institutional context on the performance of RRI pillars.

6. The Weaker the Institutional Context, the Harder the Effective Participation and Application of the RRI Approach

Following the previous section, institutions play an essential role in coordinating the innovation process since they represent the local context's crystallization. At the same time, institutions act as an approximation for understanding the values and behaviours behind how people interact and work as a "locus" or as an "action arena" that legitimizes the outcomes of such interactions. Therefore, the quality of the institutions is essential since it controls the different ways that stakeholders interact, different levels of effective participation and, thus, impacts the ability of members of the local community (researches, entrepreneurs, policymakers, NGOs) to perform RRI activities to achieve responsible innovation process effectively.

Strong institutional contexts facilitate the effective participation of stakeholders. In such cases, RRI guidelines, which demands freedom of speech, gender equality, power balance, among other attributes to perform effective participation, are easier to apply, and the results from the participatory process can be interpreted as an indication/good representation of how participative the innovation process is. On the other hand, the effects are the opposite in the case of weak institutions when the participatory process required by the RRI framework can be jeopardized by a lack of coordination in the process of engagement, an imbalanced discussion impeding the free access to information and freedom to speak out [65].

It is the case for most biofuel innovation implementation processes since biofuel production often occurs in rural areas in developing countries, far from the large urban centers where the developers of the technology or the framework's creators are found. It highlights the inherent difficulties in the implementation phase in regions where the culture of participation, history, and economic context are very different from the Western developed countries, where these contextual RRI elements are largely accepted and implemented through traditional and robust institutions. Brazil is an interesting case since it is considered technologically developed and with good institutions. However, due economic history, cultural values, and deep inequality context, the implementation phases are often done in top-down manner and therefore, is not aligned with the participative process as required by the RRI framework. Even though this lack of participation in the implementation process of biofuels projects, the outcomes of these projects for the local communities seem to point to inclusive impacts [66]. Timmermans [6] described the RRI approach as still a "predominantly Western affair". The absence of credible and stable institutions can jeopardize the free and effective participation of less privileged people who may not have full knowledge of their rights or access to information and, consequently, may have a limited capacity to analyze innovation's potential issues.

In a case study carried out in Tanzania described by Balkema and Pols [67], Jatropha's investments were stimulated by the 2003 EU directive to promote biofuels. However, the authors noted that small biofuel producers were left behind in the discussions on Jatropha's expansion. The "exclusion" of this vital stakeholder has led to missed opportunities. Considering this "weak inclusion" or failure to assure "effective participation" of the right set of agents, the authors claim that the EU failed to anticipate the impacts of its biofuel policy and did not include small farmers' voices.

There was a shift in expectations regarding biofuel production from Jatropha from the large-scale cultivation of a "miracle culture" that could provide sustainable biodiesel for Europe to a "biofuel niche culture". It turns to be viewed as a crop with relatively low yields, suitable for bio-oil production for local use. The authors conclude that the impacts of large-scale production of Jatropha have contributed to a series of problems that conflict with sustainability ideals, such as food insecurity (through land-use changes that negatively impacted food planting), the marginalization of small farmer participation as an actor of interest, loss of soil fertility and biodiversity loss. They, therefore, concluded that the case has proven to be unsustainable and irresponsible.

However, the article did not analyze the role of Tanzania's historical and economic context. How can an external institution such as the EU, with such a major difference in values and culture to an Eastern country, be responsible for identifying the legitimate stakeholders who should sit at the negotiation table? To what extent did small-scale farmers trust the process and the mediators? Were these farmers able to make their voices and concerns heard? It seems that the lack of a local institution as coordinator, with a better understanding of Tanzania's context and traditions, has jeopardized the whole RRI process, failing to establish an adequate set of engaged stakeholders.

Therefore, what can be concluded from the use of RRI in this case study is the failure to achieve effective participation. These failures led to a poorly implemented innovation with negative impacts. The "irresponsible character of the innovation" is connected with the implementation process and not with the biofuels innovation's direct material impacts in this local context. Indeed, we could judge biofuels as irresponsible if the framework had specific impact criteria to do so. For example, about food security (total food imported versus locally produced), loss of soil fertility (percentage of fertilization need compared to the past), and biodiversity loss (percentage of deforestation). Although RRI is about meeting grand challenges and aligning with society's values, the framework does not have these direct material criteria. Material issues as food security, loss of fertility, biodiversity loss, and others can appear if, and only if, the chosen set of invited stakeholders effectively participate in the discussions and bring those themes to the negotiation table. Instead, the RRI as a normative processual framework has the criteria of inclusiveness, which were done without considering institutions. Then, its results are not enough elements to judge the impacts of the innovation itself as irresponsible.

In another study on India, De Hoop, Pols, and Romijn [44] studied a Jatropha project carried out by the University of Agricultural Sciences in Bangalore (an entrepreneurial actor with an academic profile). One of the project's main goals was to improve the natural environment's quality and the lives of the people involved (inclusion as an outcome). In practice, however, several problems were highlighted due to the lack of proper inclusion of local producers in the project's design (inclusion of process or effective participation). Although the authors mention that the process respected the inclusion pillar, they expressly pointed out the low quality of involvement, which meant that farmers failed to adhere to the project due to restrictions on water access for irrigation (failure in participation implied failures in anticipation process activities). Both the farmers and the researchers had to deal with power structures. The local institutions responsible for carrying out the process were not strong enough to mediate these power differences or effectively include a range of stakeholders and guarantee their right to free speech (failure in responsiveness). The authors concluded that, although the project did not negatively impact the local population and the environment, it also did not result in the expected positive impacts. The authors present evidence that the project did not sufficiently meet the four dimensions of RRI. Therefore, it concluded that the innovation project should not be implemented since RRI is about 'innovating responsibly—or not innovating at all' [44] (p. 129).

However, our interpretation of the proposal of Stilgoe et al. [43] together with other authors on institutional theory such as Ostrom [56] and Acemoglu and Robinson [60] leads us to conclude otherwise. We understand that due to the lack of effective local actors' participation, the inclusion, anticipation, reflexivity, and responsiveness dimensions were jeopardized. There was no vital institution to lead the project that impaired the RRI dimensions, and idiosyncrasies emerged. For example, should a university that does not have an entrepreneurial mindset be responsible for carrying out a "business enterprise" with all inherent conflicts of interest of these roles? Should universities be responsible for advocating biofuel policy or negotiating trade agreements with biofuel dealers? Would the recommendation of "abandoning the innovation process", as proposed by the authors, be considered a well-thought-out and fair decision for the local community? Would that put the potential positive impact of the venture at risk just because it did not meet the expectations of participation included in the assessment tool created according to values different from the local culture? Again, the question is posed: should this conclusion be interpreted as a failure on the part of the biofuels project in India or the failure to adhere to, adapt, and implement the core RRI concepts? The analysis must first examine contextual elements of governance and local institutions in greater depth to be considered valid alternatives. After all, is it better to achieve some benefits or no benefits at all? The academic community must address this ethical question.

The two cases above did not consider the local contexts in their analyses to interpret the "possible participation process" to be implemented. As a result, we understand their conclusions reflected more a lack of adherence or adequacy of the RRI process to those contexts than the "irresponsible character of the innovations developed" in terms of its potential impacts. Both cases make conclusions about the final impacts of the innovation (calling them irresponsible or suggesting to give up the innovation) using an assessment tool (RRI) that focuses mainly on its development's participatory process. When the participation is not ideal according to RRI requirements, instead of analysing the role of institutions that coordinate the process, the author criticizes the innovation itself's potential impacts.

In contrast, reflections on the RRI approach's limits and claims for a local context analysis through an institutional approach were echoed in several studies. Chaturvedi et al. [68], for example, analyze governments' role in defining the pathways to develop innovations in the transgenic cotton sector. The authors highlighted the lack of transparency (China) and credibility (India) as evidence of weak institutions that jeopardize the RRI process and responsible innovation. Hartley et al. [69], from a different angle, analyzed the difficulties in implementing low-tech innovations using the RRI approach in transcontinental projects (Europe and Africa) and, even when the research team considered the approach, the "inclusion pillar" reflected "a process of discursive exclusion." In the case of Vietnam, Voeten and Naudé [70] analyzed the impact on weaker institutions' responsiveness dimension previously coopted by entrepreneurs. They were unable to coordinate the process to reap benefits for society as a whole.

In Latín América, de Campos et al. [71] defend an interconnection between the notion of political accountability (one aspect of responsiveness) and institutions, since the difficulty in predicting the trajectory of innovation and the absence of a deliberation arena can strongly impact the innovation trajectory of a vaccine for the dengue mosquito. Macnaghten et al. [45] and Vasen [65] explore the difficulties and limitations of employing the RRI framework in contexts with different capabilities, values, and institutions, and the risk that this can be perceived as a new model of intellectual neo-colonialism.

Figure 3 represents how the complementary analyses of institutions can help define strategies to implement the RRI approach to achieve the effective participation required and how to analyze the results of the RRI application when the local institutions are not capable of achieving an effective participatory process.

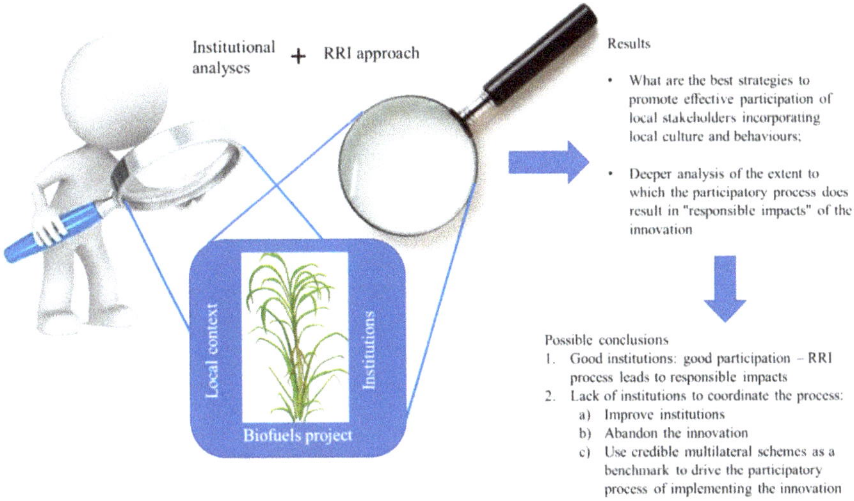

Figure 3. Complementing Responsible Research and Innovation Framework (RRI) analysis (created by the authors).

A preliminary institutional analysis is required to understand actors' ability in different contexts to implement effective participation. When local institutions can guarantee effective participation, the four pillars of RRI will probably achieve their aims. The participative process tends to generate positive impacts of the innovation, achieving the responsible innovation's ultimate goal.

However, when the institutional framework is not capable of coordinating the process (lack of trustable organization or properly enforcement power, for example), the institutional analysis helps to explore alternative strategies to improve local stakeholders' participation, taking into account their culture, value, beliefs and ways of interacting. In such cases, three possibilities are listed below, based on the literature referred to in the previous sections and the author's reflections:

(1) Improve institutions. As stressed by Taebi et al. [52] and Acemoglu and Robinson [60], this alternative is a lengthy process and one that requires strong political capabilities with solid social movements. Although it is the most laudable and sustainable alternative in the long run and hence it should be strived for continuously, it is not an option for some projects since institutional strengthening requires a longer timeframe than the innovation processes usually addressed by the RRI. It is possible that, in some cases of grassroots innovation (one which is raised from the ground or by the community), it may be an option.

(2) Abandon innovation. The second option, the most radical, should not be considered without in-depth institutional analysis. As discussed above, abandoning an innovation because the effective participation required in the RRI pillars are not strictly followed would impact communities and hinder the local development process. Would this effect be ethical and fair to local communities, just because the innovation process did not match the mindset of a framework designed externally? This option seems to be applied when the institutional analysis has demonstrated a lack of institutions that can minimally guarantee to use the RRI principles, therefore showing that no participatory process is possible. Even in this case however, the impacts of the innovation can be positive in amplifying the access to goods and services for a larger part of the population. Should such kind of innovation be called irresponsible because of the lack of a participative development process? Or its impact is what defines the responsible character of the innovation?

(3) Using alternative external schemes or standards as a trustable guide for a more participative implementation process. For example, the Sustainable Development Goal (SDG) or the IFC Standards, Principles of Responsible Investment, and other sectorial schemes aim to assure

an inclusive and participatory process as principles of action. From this, it is possible to define strategies to guarantee the most participative process as possible (thus, responsible through the RRI lens and given that specific context) and, as a result, the accomplishment of positive outcomes in terms of inclusion, anticipation, reflexivity, and responsiveness.

It is important to emphasize that the three options above are alternatives to improve participation during the innovation development and implementation phases and raise the chances to achieve the positive impacts of the innovation. None of the options scanned guarantees that participation alone will produce outcomes aligned with society's needs. It depends on quality, diversity, critical skills, access to information, and other factors. It is precisely the problem of considering the framework to assess the responsible character of the Innovation. RRI is useful to inspire a participative process but what can be evaluated through the RRI approach is the process of participation and engagement, and not the innovation impact itself.

7. Limitations

Although mentioned in the literature review and discussion among our peers, we intentionally do not describe the contours of "strong" or good institutions because we do not want to repeat the mistake of crystallizing a vision of institutions that do not consider the local contexts that are naturally specific. The same reasoning is applied to the definition of the 'minimum requirements' for a trustable private sector organization to coordinate the development process when the institutions are not strong.

8. Conclusions

Worldwide biofuel expansion is essential if we are to transition to a biobased economy. In this sense, emerging technologies cannot disregard societies' social and environmental demands in which the innovations are implemented. Several frameworks emerged to assess how sustainable is the innovation in question, RRI among them.

To address the "responsible character of the innovation" as framed by RRI, it is important to understand what "responsible innovation" means. Is it the innovation that counted on stakeholders' effective participation during its development process or the innovation whose impacts are aligned with society's aims and needs? RRI framework seems to incorporate the premise that effective participation (as extracted from the good performance of the four pillars) will guarantee responsible impacts as defined by the participants and, therefore, the alignment of the innovation with the society needs.

However, there are three problems with this premise:

(1) It requires a quality of the participatory process dependent on culture, values, and behaviours that vary among the countries and regions. What makes it difficult to define upfront processual indicators that could assess effective participation quality in any context.

(2) Regarding the material elements that would generate indicators, metrics, and measurements to assess the responsible innovation impacts, those elements are dependent on the quality of the participation process. They cannot be upfront defined as common material metrics and indicators.

(3) Even when the process dimension of the innovation development in RRI does consider those contextual elements (culture and values) as fundamental requirements to perform effective participation, nothing can be predicted about how positive are the innovation's impacts. The improvements over the process of participation cannot automatically guarantee the positive impacts of the innovation.

Through the analysis of the literature about different case studies, we identified that conclusions about the responsible character of innovation that uses the RRI as an assessment tool usually reflect more the difficulties in applying the framework (the ideal participatory process in regions with a different culture, types of institutions and values) than the real "responsible" impact of the innovation in the study. RRI is a normative framework designed to influence the process of innovation and not to define the quality of the innovation's impacts.

We then conclude that using the RRI framework in regions of biofuels implementation projects reflects more the difficulties involved in applying guidelines that were, perhaps unconsciously, embedded with western values than the responsible character of the innovation impacts, which usually are defined upfront materially and measurably. Thus, RRI cannot be used as an assessment tool since it does not have the material metrics to measure how responsible or positive or negative the impacts of the innovation are implemented.

Irrespective of the critics, RRI remains a necessary normative framework to inspire and guide responsible innovation through the participatory process (the core element to the effectiveness of anticipation, reflexiveness, responsiveness, and inclusion) and achieve the innovation whose impacts are good for society. Its application should complement an initial analysis of institutions' quality to encapsulate local context and values. Understanding the institutional context before RRI application can help map the limitations and gaps and better define strategies to improve stakeholders' participation in the four pillars of the RRI framework and, consequently, raise the chances of promising outcomes regarding inclusive impacts of the innovation.

Future research should be carried out to understand whether innovation can be considered as responsible when substantive inclusion or effective participation is not entirely possible; whether it is possible to achieve inclusion as an outcome of a biofuel implementation project without proper participation (inclusion of process) in the innovation process, and how inclusive has the biofuel expansion been in different contexts.

Author Contributions: A.M.P., conceptualization, methodology, software, formal analysis, investigation, writing—original draft and review and editing, visualization; G.B., conceptualization, formal analysis, writing—original draft preparation, writing—review and editing; M.P.P. conceptualization, writing—review and editing; L.A., conceptualization, writing—review and editing, resources, supervision; P.O. conceptualization, writing—review and editing, supervision, validation, project administration, funding acquisition; J.M.F.J.D.S., conceptualization, validation, supervision, project administration, funding acquisition. All authors have read and agreed to the published version of the manuscript.

Funding: This study was financed in part by the Coordenação de Aperfeiçoamento de Pessoal de Nível Superior—Brasil (CAPES), finance Code 001 and by FAPESP—the São Paulo Research Foundation 2013/50950-5.

Conflicts of Interest: No potential conflict of interest was reported by the authors.

References

1. Blok, V.; Hoffmans, L.; Wubben, E.F.M. Stakeholder engagement for responsible innovation in the private sector: Critical issues and management practices. *J. Chain Netw. Sci.* **2015**, *15*, 147–164. [CrossRef]
2. Carbajo, R.; Cabeza, L.F. Renewable energy research and technologies through responsible research and innovation looking glass: Reflexions, theoretical approaches and contemporary discourses. *Appl. Energy* **2018**, *211*, 792–808. [CrossRef]
3. Eizagirre, A.; Rodríguez, H.; Ibarra, A. Politicizing Responsible Innovation: Responsibility as Inclusive Governance. *Int. J. Innov. Stud.* **2017**. [CrossRef]
4. Hahn, J.; Ladikas, M. Responsible Research and Innovation: a Global Perspective. *Enterp. Work Innov. Stud.* **2014**, *10*, 9–27.
5. Blok, V.; Lemmens, P. The emerging concept of responsible innovation. Three reasons why it is questionable and calls for a radical transformation of the concept of innovation. In *Responsible Innovation 2: Concepts, Approaches, and Applications*; Koops, B.-J., Oosterlaken, I., Romijn, H., Swierstra, T., van den Hoven, J., Eds.; Springer International Publishing: Cham, Switzerland, 2015; pp. 19–35. ISBN 978-3-319-17307-8.
6. Timmermans, J. Mapping the RRI Landscape: an Overview of Organizations, Projects, Persons, Areas and Topics. In *Responsible Innovation 3: A European Agenda?* Asveld, L., van Dam-Mieras, R., Swierstra, T., Lavrijssen, S., Linse, K., van den, H.J., Eds.; Springer: Cham, Switzerland, 2017; pp. 21–48. ISBN 978-3-319-64834-7.
7. IRENA Trends in Renewable Energy. Available online: https://public.tableau.com/profile/irena.resource#!/vizhome/IRENARETimeSeries/Charts) (accessed on 16 July 2020).

8. Sondergaard, N. *Brazilian State-Agribusiness Relations within Global Processes of Regulatory Formation below the Corporate Food Regime*; University of Brasilia: Brasilia, Brazil, 2018; Available online: https://repositorio.unb.br/handle/10482/34516 (accessed on 15 December 2020).
9. Bórawski, P.; Bełdycka-Bórawska, A.; Szymańska, E.J.; Jankowski, K.J.; Dubis, B.; Dunn, J.W. Development of renewable energy sources market and biofuels in The European Union. *J. Clean. Prod.* **2019**, *228*, 467–484. [CrossRef]
10. Frankfurt School–UNEP. Global Trends Renewable Energy 2020. 2020. Available online: https://www.fs-unep-centre.org/global-trends-in-renewable-energy-investment-2020/ (accessed on 14 December 2020).
11. Proskurina, S.; Junginger, M.; Heinimö, J.; Tekinel, B.; Vakkilainen, E. Global biomass trade for energy—Part 2: Production and trade streams of wood pellets, liquid biofuels, charcoal, industrial roundwood and emerging energy biomass. *Biofuels Bioprod. Biorefining* **2019**, *13*, 371–387. [CrossRef]
12. Borras, S.M.; McMichael, P.; Scoones, I. The politics of biofuels, land and agrarian change: Editors' introduction. *J. Peasant Stud.* **2010**, *37*, 575–592. [CrossRef]
13. Miyake, S.; Renouf, M.; Peterson, A.; Mcalpine, C.; Smith, C. Land-use and environmental pressures resulting from current and future bioenergy crop expansion: A review. *J. Rural Stud.* **2012**, *28*, 650–658. [CrossRef]
14. Nassar, A.M.; Harfuch, L.; Bachion, L.C.; Moreira, M.R. Biofuels and land-use changes: Searching for the top model. *Interface Focus* **2011**, 224–232. [CrossRef]
15. Schlesinger, S. *Biofuels: Energy Won't Feed the Hungry*; ActionAid Brazil: Rio de Janeiro, Brazil, 2014; Available online: https://actionaid.org/publications/2014/biofuels-energy-wont-feed-hungry (accessed on 15 December 2020).
16. Escobar, J.C.; Lora, E.S.; Venturini, O.J.; Yáñez, E.E.; Castillo, E.F.; Almazan, O. Biofuels: Environment, technology and food security. *Renew. Sustain. Energy Rev.* **2009**, *13*, 1275–1287. [CrossRef]
17. Richard, T. Ask the Experts: The food versus fuel debate. *Biofues* **2012**, *3*, 635–648.
18. Rosillo-Calle, F. Food versus Fuel: Toward a New Paradigm—The Need for a Holistic Approach. *ISRN Renew. Energy* **2012**, *2012*, 1–15. [CrossRef]
19. Mol, A.P.J. Boundless biofuels? Between environmental sustainability and vulnerability. *Sociol. Ruralis* **2007**, *47*, 297–315. [CrossRef]
20. Gallardo, A.L.C.F.; Bond, A. Capturing the implications of land use change in Brazil through environmental assessment: Time for a strategic approach? *Environ. Impact Assess. Rev.* **2011**, *31*, 261–270. [CrossRef]
21. Ribeiro, B.E. Beyond commonplace biofuels: Social aspects of ethanol. *Energy Policy* **2013**, *57*, 355–362. [CrossRef]
22. Filoso, S.; Do Carmo, J.B.; Mardegan, S.F.; Lins, S.R.M.; Gomes, T.F.; Martinelli, L.A. Reassessing the environmental impacts of sugarcane ethanol production in Brazil to help meet sustainability goals. *Renew. Sustain. Energy Rev.* **2015**, *52*, 1847–1856. [CrossRef]
23. Ciervo, M.; Schmitz, S. Sustainable biofuel: A question of scale and aims. *Morav. Geogr. Rep.* **2017**, *25*, 220–233. [CrossRef]
24. Martinelli, L.A.; Filoso, S. Balance between food production, biodiversity and ecosystem services in Brazil: A challenge and an opportunity. *Biota Neotrop.* **2009**, *9*, 21–25. [CrossRef]
25. Wilkinson, J.; Herrera, S. Biofuels in brazil: Debates and impacts. *J. Peasant Stud.* **2010**, *37*, 749–768. [CrossRef]
26. Pilgrim, S.; Harvey, M. Battles over biofuels in Europe: NGOs and the politics of markets. *Sociol. Res. Online* **2010**, *15*. [CrossRef]
27. Tilman, D.; Socolow, R.; Foley, J.A.; Hill, J.; Larson, E.; Lynd, L.; Pacala, S.; Reilly, J.; Searchinger, T.; Somerville, C.; et al. Beneficial Biofuels—The Food, Energy, and Environment Trilemma. *Science* **2009**, *325*, 270–271. [CrossRef] [PubMed]
28. Harvey, M.; Pilgrim, S. The new competition for land: Food, energy, and climate change. *Food Policy* **2011**, *36*, 1–12. [CrossRef]
29. Assato, M.M.; de Moares, M.A.F.; de Oliveira, F.C.R. Impactos sócio-econômicos da expansão do setor bioenergético no estado do Mato Grosso do Sul: Os casos dos municípios de Nova Alvorada do Sul e Rio Brilhante. In Proceedings of the 49o Congresso da Sober, Belo Horizonte, Brazil, 24–27 July 2011; Available online: https://bv.fapesp.br/pt/auxilios/78723/49o-congresso-da-sociedade-brasileira-de-economia-administracao-e-sociologia-rural-sober/ (accessed on 15 December 2020).
30. Bacchi, M.R.P.; Caldarelli, C.E. Impactos socioeconômicos da expansão do setor sucroenergético no Estado de São Paulo, entre 2005 e 2009. *Nov. Econ.* **2015**, *25*, 218. [CrossRef]

31. Diaz-Chavez, R.A. Assessing biofuels: Aiming for sustainable development or complying with the market? *Energy Policy* **2011**, *39*, 5763–5769. [CrossRef]
32. van Eijck, J.; Faaij, A.P.C. *Socio-Economic Impacts of Bioenergy Production*; Rutz, D., Janssen, R., Eds.; Springer International Publishing: Cham, Switzerland, 2014; ISBN 978-3-319-03828-5.
33. Moraes, M.A.F.D.; Oliveira, F.C.R.; Diaz-Chavez, R.A. Socio-economic impacts of Brazilian sugarcane industry. *Environ. Dev.* **2015**, *16*, 31–43. [CrossRef]
34. Rutz, D.; Janssen, R. Socio-economic impacts of biofuels on land use change. In *Socio-Economic Impacts of Bioenergy Production*; Biomass, D.R., Janssen, R., Eds.; Springer: Munich, Germany, 2014; Volume 9783319038, pp. 81–93. ISBN 978-3-319-03829-2.
35. GNESD. Bioenergy: The Potential for Rural Development and Poverty Alleviation. Global Network on Energy for Sustainable Development (GNESD). Summary for Policy-Makers. 2011. Available online: https://orbit.dtu.dk/en/publications/bioenergy-the-potential-for-rural-development-and-poverty-allevia (accessed on 15 December 2020).
36. Creutzig, F.; Corbera, E.; Bolwig, S.; Hunsberger, C. Integrating place-specific livelihood and equity outcomes into global assessments of bioenergy deployment. *Environ. Res. Lett.* **2013**, *8*. [CrossRef]
37. Souza, L.G.A.; Moraes, M.A.F.D.; Dal Poz, M.E.S.; Silveira, J.M.F.J. Collaborative Networks as a measure of the Innovation Systems in second-generation ethanol. *Scientometrics* **2015**, *103*, 355–372. [CrossRef]
38. Bueno, C.D.S.; Silveira, J.M.F.J.; Buainain, A.M. Innovation, networks and the paradigm of biofuels. *Int. J. Entrep. Small Bus.* **2018**, *35*, 452–484. [CrossRef]
39. Pashaei Kamali, F.; Borges, J.A.R.; Osseweijer, P.; Posada, J.A. Towards social sustainability: Screening potential social and governance issues for biojet fuel supply chains in Brazil. *Renew. Sustain. Energy Rev.* **2018**, *92*, 50–61. [CrossRef]
40. Mousavi, S.; Bossink, B. Corporate-NGO partnership for environmentally sustainable innovation: Lessons from a cross-sector collaboration in aviation biofuels. *Environ. Innov. Soc. Transit.* **2020**, *34*, 80–95. [CrossRef]
41. Fortin, E. Transnational multi-stakeholder sustainability standards and biofuels: Understanding standards processes. *J. Peasant Stud.* **2013**, *40*, 563–587. [CrossRef]
42. Owen, R.; Macnaghten, P.; Stilgoe, J. Responsible research and innovation: From science in society to science for society, with society. *Sci. Public Policy* **2012**, *39*, 751–760. [CrossRef]
43. Stilgoe, J.; Owen, R.; Macnaghten, P. Developing a framework for responsible innovation. *Res. Policy* **2013**, *42*, 1568–1580. [CrossRef]
44. De Hoop, E.; Pols, A.; Romijn, H. Limits to responsible innovation. *J. Responsible Innov.* **2016**, *3*, 110–134. [CrossRef]
45. Macnaghten, P.; Owen, R.; Stilgoe, J.; Wynne, B.; Azevedo, A.; de Campos, A.; Chilvers, J.; Dagnino, R.; di Giulio, G.; Frow, E.; et al. Responsible innovation across borders: Tensions, paradoxes and possibilities. *J. Responsible Innov.* **2014**, *1*, 191–199. [CrossRef]
46. Asveld, L.; Ganzevles, J.; Osseweijer, P. Trustworthiness and Responsible Research and Innovation: The Case of the Bio-Economy. *J. Agric. Environ. Ethics* **2015**, *28*, 571–588. [CrossRef]
47. Asveld, L.; van Dam-Mieras, R. Introduction: Responsible Research and Innovation for Sustainability. In *Responsible Innovation 3*; Asveld, L., van Dam-Mieras, R., Swierstra, T., Lavrijssen, S., Linse, K., van den Hoven, J., Eds.; Springer International Publishing: Cham, Switzerland, 2017; pp. 1–6. ISBN 978-3-319-64834-7.
48. Wong, P.H. Responsible innovation for decent nonliberal peoples: a dilemma? *J. Responsible Innov.* **2016**, *3*, 154–168. [CrossRef]
49. Bronson, K. Diversity by design: Substantive inclusion for transformative agricultural value chains. In *Proceedings of the IBIS Seminar—Building Inclusive Agricultural Value Chains Online Seminar Series*; Biotechnology & Society Section at TU Delft, Business Management & Organisation Division: Delft, The Netherlands, October 2020.
50. Noorman, M.; Swierstra, T.; Zandbergen, D. Questioning the Normative Core of RI: The Challenges Posed to Stakeholder Engagement in a Corporate Setting. In *Responsible Innovation 3*; Springer International Publishing: Cham, Switzerland, 2017; pp. 231–249. ISBN 978-3-319-64834-7.
51. Stober, D.; Suškevičs, M.; Eiter, S.; Müller, S.; Martinát, S.; Buchecker, M. What is the quality of participatory renewable energy planning in Europe? A comparative analysis of innovative practices in 25 projects. *Energy Res. Soc. Sci.* **2021**, *71*. [CrossRef]

52. Taebi, B.; Correljé, A.; Cuppen, E.; Dignum, M.; Pesch, U. Responsible innovation as an endorsement of public values: The need for interdisciplinary research. *J. Responsible Innov.* **2014**, *1*, 118–124. [CrossRef]
53. Callon, M.; Lascoumes, P.; Barthe, Y. *Acting in an Uncertain World—An Essay on Technical Democracy*; The MIT Press: Cambridge, MA, USA, 2009; ISBN 978-0-262-03382-4.
54. Pansera, M.; Owen, R. Framing inclusive innovation within the discourse of development: Insights from case studies in India. *Res. Policy* **2018**, *47*, 23–34. [CrossRef]
55. Setiawan, A.D.; Singh, R. Responsible innovation in practice: The adoption of solar PV in telecom towers in Indonesia. In *Responsible Innovation 2: Concepts, Approaches, and Applications*; Springer International Publishing: Cham, Switzerland, 2015; pp. 225–244. ISBN 9783319173085.
56. Ostrom, E. *Designing Complexity to Govern Complexity*; World Bank: Washington, DC, USA, 1995; ISBN 0821334158.
57. Wells, P.; Zapata, C. Eco-industrialism: The Potential for Inclusive Growth with Bio-Plastic Production in Brazil Using Sugarcane Ethanol. In Proceedings of the 3rd International Workshop: Advances in Cleaner Production, São Paulo, Brazil, 18–20 May 2011.
58. Feinberg, R.; Stein, E.; Tommasi, M.; Echebarría, K.; Lora, E.; Payne, M. The Politics of Policies: Economic and Social Progress in Latin America, 2006 Report. *Foreign Aff.* **2006**, *85*, 169. [CrossRef]
59. Barney, J.B.; Nelson, R.R.; Winter, S.G. *An Evolutionary Theory of Economic Change*; Havard University Press: Cambridge, MA, USA, 1987; Volume 32, ISBN 0674272285.
60. Acemoglu, D.; Robinson, J. *Why Nations Fail: The Origins of Power, Prosperity and Poverty*; Business, C., Ed.; Elsevier Ltd.: Washington, DC, USA, 2012; ISBN 978-0307719219.
61. Ostrom, E. *Understanding Institutional Diversity*, 2005th ed.; Princeton University Press: Princeton, NJ, USA, 2005; Volume 132, ISBN 9780691122076.
62. Wickson, F.; Carew, A.L. Quality criteria and indicators for responsible research and innovation: Learning from transdisciplinarity. *J. Responsible Innov.* **2014**, *1*, 254–273. [CrossRef]
63. Levitsky, S.; Murillo, M.V. Building Institutions on Weak Foundations. *J. Democr.* **2013**, *24*, 93–107. [CrossRef]
64. Randles, S.; Youtie, J.; Guston, D.H.; Shapira, P.; Wickson, F.; Rip, A. A Transatlantic Conversation on Responsible Innovation and Responsible Governance. In Proceedings of the Annual Meeting of the Society for the Study of Nanocience and Emerging Technologies (S.NET), Enschede, The Netherlands, 22–25 October 2012.
65. Vasen, F. Responsible Innovation in Developing Countries: An Enlarged Agenda. In *Responsible Innovation 3*; Springer International Publishing: Cham, Switzerland, 2017; pp. 93–109. ISBN 9783319648347.
66. Marques Postal, A.; Pashaei Kamali, F.; Asveld, L.; Osseweijer, P.; da Silveira, J.M.F.J. The impact of sugarcane expansion in Brazil: Local stakeholders' perceptions. *J. Rural Stud.* **2020**, *73*, 147–162. [CrossRef]
67. Balkema, A.; Pols, A. Biofuels: Sustainable Innovation or Gold Rusch? Identifying Responsibilities form Biofuels Innovations. *J. Responsible Innov.* **2014**, *56*, 283–303. [CrossRef]
68. Chaturvedi, S.; Ravi Srinivas, K.; Kumar, A. Agriculture technology choices and the responsible research and innovation (RRI) framework: Emerging experiences from China and India. *Asian Biotechnol. Dev. Rev.* **2016**, *18*, 93–111.
69. Hartley, S.; McLeod, C.; Clifford, M.; Jewitt, S.; Ray, C. A retrospective analysis of responsible innovation for low-technology innovation in the Global South. *J. Responsible Innov.* **2019**, *6*, 143–162. [CrossRef]
70. Voeten, J.J.; Naudé, W.A. Regulating the negative externalities of enterprise cluster innovations: Lessons from Vietnam. *Innov. Dev.* **2014**, *4*, 203–219. [CrossRef]
71. De Campos, A.S.; Hartley, S.; de Koning, C.; Lezaun, J.; Velho, L. Responsible Innovation and political accountability: Genetically modified mosquitoes in Brazil. *J. Responsible Innov.* **2017**, *4*, 5–23. [CrossRef]

MDPI
St. Alban-Anlage 66
4052 Basel
Switzerland
Tel. +41 61 683 77 34
Fax +41 61 302 89 18
www.mdpi.com

Sustainability Editorial Office
E-mail: sustainability@mdpi.com
www.mdpi.com/journal/sustainability

www.ingramcontent.com/pod-product-compliance
Lightning Source LLC
Chambersburg PA
CBHW062101220926
51850CB00010B/2851
* 9 7 8 3 0 3 6 5 0 4 6 8 1 *